ALEKS Math in 10 Days!

The Most Effective ALEKS Math Crash Course

By

Reza Nazari

All inquiries should be addressed to:
info@effortlessMath.com
www.EffortlessMath.com

ISBN: 978-1-63719-249-8

Published by: **Effortless Math Education Inc.**

For Online Math Practice Visit www.EffortlessMath.com

Welcome to
ALEKS Math Prep
2023

Thank you for choosing Effortless Math for your ALEKS Math test preparation and congratulations on making the decision to take the ALEKS test! It's a remarkable move you are taking, one that shouldn't be diminished in any capacity. That's why you need to use every tool possible to ensure you succeed on the test with the highest possible score, and this extensive study guide is one such tool.

If math has never been a strong subject for you, don't worry! This book will help you prepare for (and even ACE) the ALEKS Math assessment. As test day draws nearer, effective preparation becomes increasingly more important. Thankfully, you have this quick study guide to help you get ready for the test. With this guide, you can feel confident that you will be more than ready for the ALEKS Math test when the time comes.

First and foremost, it is important to note that this book is a quick study guide and not a textbook. It is best read from cover to cover. Every lesson of this "self-guided math book" was carefully developed to ensure that you are making the most effective use of your time while preparing for the test. This up-to-date guide reflects the 2023 test guidelines and will put you on the right track to hone your math skills, overcome exam anxiety, and boost your confidence, so that you can have your best to succeed on the ALEKS Math test.

This study guide will:

☑ Explain the format of the ALEKS Math test.

☑ Describe specific test-taking strategies that you can use on the test.

☑ Provide ALEKS Math test-taking tips.

☑ Review all ALEKS Math concepts and topics you will be tested on.

☑ Help you identify the areas in which you need to concentrate your study time.

☑ Offer exercises that help you develop the basic math skills you will learn in each section.

☑ Give **2 realistic and full-length practice tests** (featuring new question types) with detailed answers to help you measure your exam readiness and build confidence.

This resource contains everything you will ever need to succeed on the ALEKS Math test. You'll get in-depth instructions on every math topic as well as tips and techniques on how to answer each question type. You'll also get plenty of practice questions to boost your test-taking confidence.

In addition, in the following pages you'll find:

➢ **How to Use This Book Effectively** – This section provides you with step-by-step instructions on how to get the most out of this comprehensive study guide.

How to study for the ALEKS Math Test – A six-step study program has been developed to help you make the best use of this book and prepare for your ALEKS Math test. Here you'll find tips and strategies to guide your study program and help you understand ALEKS Math and how to ace the test.

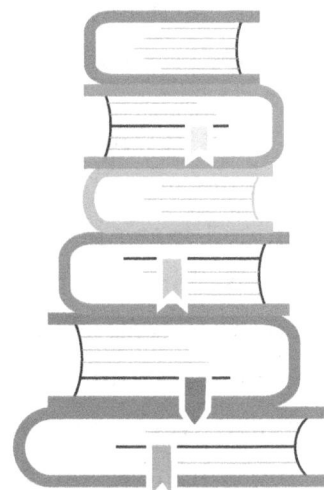

➢ **ALEKS Math Review** – Learn everything you need to know about the ALEKS Math test.

➢ **ALEKS Math Test-Taking Strategies** – Learn how to effectively put these recommended test-taking techniques into use for improving your ALEKS Math score.

➢ **Test Day Tips** – Review these tips to make sure you will do your best when the big day comes.

Effortless Math's ALEKS Online Center

Effortless Math Online ALEKS Center offers a complete study program, including the following:

✓ Step-by-step instructions on how to prepare for the ALEKS Math test

✓ Numerous ALEKS Math worksheets to help you measure your math skills

✓ Complete list of ALEKS Math formulas

✓ Video lessons for all ALEKS Math topics

✓ Full-length ALEKS Math practice tests

✓ And much more…

No Registration Required.

Visit **EffortlessMath.com/ALEKS** to find your online ALEKS Math resources.

How to Use This Book Effectively

Look no further when you need a study guide to improve your math skills to succeed on the math portion of the ALEKS test. Each chapter of this comprehensive guide to the ALEKS Math will provide you with the knowledge, tools, and understanding needed for every topic covered on the test.

It's imperative that you understand each topic before moving onto another one, as that's the way to guarantee your success. Each chapter provides you with examples and a step-by-step guide of every concept to better understand the content that will be on the test. To get the best possible results from this book:

➢ **Begin studying long before your test date**. This provides you ample time to learn the different math concepts. The earlier you begin studying for the test, the sharper your skills will be. Do not procrastinate! Provide yourself with plenty of time to learn the concepts and feel comfortable that you understand them when your test date arrives.

➢ **Practice consistently**. Study ALEKS Math concepts at least 20 to 30 minutes a day. Remember, slow and steady wins the race, which can be applied to preparing for the ALEKS Math test. Instead of cramming to tackle everything at once, be patient and learn the math topics in short bursts.

➢ Whenever you get a math problem wrong, **mark it off, and review it later** to make sure you understand the concept.

➢ Start each session by looking over the previous related material.

➢ Once you've reviewed the book's lessons, **take a practice test** at the back of the book to gauge your level of readiness. Then, review your results. Read detailed answers and solutions for each question you missed.

➢ **Take another practice test** to get an idea of how ready you are to take the actual exam. Taking the practice tests will give you the confidence you need on test day. Simulate the ALEKS testing environment by sitting in a quiet room free from distraction. Make sure to clock yourself with a timer.

How to Study for the ALEKS Math Test

Studying for the ALEKS Math test can be a really daunting and boring task. What's the best way to go about it? Is there a certain study method that works better than others? Well, studying for the ALEKS Math can be done effectively. The following six-step program has been designed to make preparing for the ALEKS Math test more efficient and less overwhelming.

Step **1** - Create a study plan
Step **2** - Choose your study resources
Step **3** - Review, Learn, Practice
Step **4** - Learn and practice test-taking strategies
Step **5** - Learn the ALEKS Test format and take practice tests
Step **6** - Analyze your performance

STEP 1: Create a Study Plan

It's always easier to get things done when you have a plan. Creating a study plan for the ALEKS Math test can help you to stay on track with your studies. It's important to sit down and prepare a study plan with what works with your life, work, and any other obligations you may have. Devote enough time each day to studying. It's also a great idea to break down each section of the exam into blocks and study one concept at a time.

It's important to understand that there is no "right" way to create a study plan. Your study plan will be personalized based on your specific needs and learning style.

Follow these guidelines to create an effective study plan for your ALEKS Math test:

★ **Analyze your learning style and study habits** – Everyone has a different learning style. It is essential to embrace your individuality and the unique way you learn. Think about what works and what doesn't work for you. Do you prefer ALEKS Math prep books or a combination of textbooks and video lessons? Does it work better for you if you study every night for thirty minutes or is it more effective to study in the morning before going to work?

★ **Evaluate your schedule** – Review your current schedule and find out how much time you can consistently devote to ALEKS Math study.

★ **Develop a schedule** – Now it's time to add your study schedule to your calendar like any other obligation. Schedule time for study, practice, and review. Plan out which topic you will study on which day to ensure that you're devoting enough time to each concept. Develop a study plan that is mindful, realistic, and flexible.

★ **Stick to your schedule** – A study plan is only effective when it is followed consistently. You should try to develop a study plan that you can follow for the length of your study program.

★ **Evaluate your study plan and adjust as needed** – Sometimes you need to adjust your plan when you have new commitments. Check in with yourself regularly to make sure that you're not falling behind in your study plan. Remember, the most important thing is sticking to your plan. Your study plan is all about helping you be more productive. If you find that your study plan is not as effective as you want, don't get discouraged. It's okay to make changes as you figure out what works best for you.

STEP 2: Choose Your Study Resources

There are numerous textbooks and online resources available for the ALEKS Math test, and it may not be clear where to begin. Don't worry! This study guide provides everything you need to fully prepare for your ALEKS Math test. In addition to the book content, you can also use Effortless Math's online resources. (video lessons, worksheets, formulas, etc.)

Simply visit EffortlessMath.com/ALEKS to find your online ALEKS Math resources.

STEP 3: Review, Learn, Practice

This ALEKS Math study guide breaks down each subject into specific skills or content areas. For instance, the percent concept is separated into different topics–percent calculation, percent increase and decrease, percent problems, etc. Use this book to help you go over all key math concepts and topics on the ALEKS Math test.

As you read each chapter, take notes or highlight the concepts you would like to go over again in the future. If you're unfamiliar with a topic or something is difficult for you, do additional research on it. For each math topic, plenty of instructions, step-by-step guides, and examples are provided to ensure you get a good grasp of the material. You can also find video lessons on the Effortless Math website for each ALEKS Math concept.

Quickly review the topics you do understand to get a brush-up of the material. Be sure to do the practice questions provided at the end of every chapter to measure your understanding of the concepts.

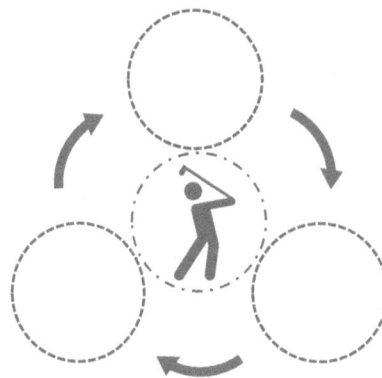

STEP 4: Learn and Practice Test-taking Strategies

In the following sections, you will find important test-taking strategies and tips that can help you earn extra points. You'll learn how to think strategically and when to guess if you don't know the answer to a question. Using ALEKS Math test-taking strategies and tips can help you raise your score and do well on the test. Apply test taking strategies on the practice tests to help you boost your confidence.

STEP 5: Learn the ALEKS Test Format and Take Practice Tests

The *ALEKS Test Review* section provides information about the structure of the ALEKS test. Read this section to learn more about the ALEKS test structure, different test sections, the number of questions in each section, and the section time limits. When you have a prior understanding of the test format and different types of ALEKS Math questions, you'll feel more confident when you take the actual exam.

Once you have read through the instructions and lessons and feel like you are ready to go – take advantage of both of the full-length ALEKS Math practice tests available in this study guide. Use the practice tests to sharpen your skills and build confidence.

The ALEKS Math practice tests offered at the end of the book are formatted similarly to the actual ALEKS Math test. When you take each practice test, try to simulate actual testing conditions. To take the practice tests, sit in a quiet space, time yourself, and work through as many of the questions as time allows. The practice tests are followed by detailed answer explanations to help you find your weak areas, learn from your mistakes, and raise your ALEKS Math score.

STEP 6: Analyze Your Performance

After taking the practice tests, look over the answer keys and explanations to learn which questions you answered correctly and which you did not. Never be discouraged if you make a few mistakes. See them as a learning opportunity. This will highlight your strengths and weaknesses.

You can use the results to determine if you need additional practice or if you are ready to take the actual ALEKS Math test.

Looking for more?

Visit EffortlessMath.com/ALEKS to find hundreds of ALEKS Math worksheets, video tutorials, practice tests, ALEKS Math formulas, and much more.

Or scan this QR code.

No Registration Required.

ALEKS Test Review

ALEKS (Assessment and Learning in Knowledge Spaces) is an artificial intelligence-based assessment tool to measure the strengths and weaknesses of a student's mathematical knowledge. ALEKS is available for a variety of subjects and courses in K-12, Higher Education, and Continuing Education. The findings of ALEKS's assessment test help to find an appropriate level for course placement. The ALEKS math placement assessment ensures students' readiness for particular math courses at colleges.

ALEKS does not use multiple-choice questions like most other standardized tests. Instead, it utilizes adaptable and easy-to-use method that mimic paper and pencil techniques. When taking the ALEKS test, a brief tutorial helps you learn how to use ALEKS answer input tools. You then begin the ALEKS Assessment. In about 30 to 45 minutes, the test measures your current content knowledge by asking 20 to 30 questions. ALEKS is a Computer Adaptive (CA) assessment. It means that each question will be chosen on the basis of answers to all the previous questions. Therefore, each set of assessment questions is unique. The ALEKS Math assessment does not allow you to use a personal calculator. But for some questions ALEKS onscreen calculator button is active and the test taker can use it.

Key Features of the ALEKS Mathematics Assessment
Some key features of the ALEKS Math assessment are:

❖ Mathematics questions on ALEKS are adaptive to identify the student's knowledge from a comprehensive standard curriculum, ranging from basic arithmetic up to precalculus, including trigonometry but not calculus.

❖ Unlike other standardized tests, the ALEKS assessment does not provide a "grade" or "raw score." Instead, ALEKS identifies which concepts the student has mastered and what topics the student needs to learn.

❖ ALEKS does not use multiple-choice questions. Instead, students need to produce authentic mathematical input.

❖ There is no time limit for taking the ALEKS Math assessment. But it usually takes 30 to 45 minutes to complete the assessment.

The ALEKS Math score is between 1 and 100 and is interpreted as a percentage correct. A higher ALEKS score indicates that the test-taker has mastered more math concepts. ALEKS Math assessment tool evaluates mastery of a comprehensive set of mathematics skills ranging from basic arithmetic up to precalculus, including trigonometry but not calculus. It will place students in classes up to Calculus.

ALEKS Math Test-Taking Strategies

Here are some test-taking strategies that you can use to maximize your performance and results on the ALEKS Math test.

#1: USE THIS APPROACH TO ANSWER EVERY ALEKS MATH QUESTION

- Review the question to identify keywords and important information.

- Translate the keywords into math operations so you can solve the problem.

- Review the answer choices. What are the differences between answer choices?

- Draw or label a diagram if needed.

- Try to find patterns.

- Find the right method to answer the question. Use straightforward math, plug in numbers, or test the answer choices (backsolving).

- Double-check your work.

#2: ANSWER EVERY ALEKS MATH QUESTION

Don't leave any fields empty! ALEKS is a Computer Adaptive (CA) assessment. Therefore, you cannot leave a question unanswered and you cannot go back to previous questions.

Even if you're unable to work out a problem, strive to answer it. Take a guess if you have to. You will not lose points by getting an answer wrong, though you may gain a point by getting it correct!

#3 : BALLPARK

A ballpark answer is a rough approximation. When we become overwhelmed by calculations and figures, we end up making silly mistakes. A decimal that is moved by one unit can change an answer from right to wrong, regardless of the number of steps that you went through to get it. That's where ballparking can play a big part.

If you think you know what the correct answer may be (even if it's just a ballpark answer), you'll usually have the ability to estimate the range of possible answers and avoid simple mistakes.

#4 : PLUGGING IN NUMBERS

"Plugging in numbers" is a strategy that can be applied to a wide range of different math problems on the ALEKS Math test. This approach is typically used to simplify a challenging question so that it is more understandable. By using the strategy carefully, you can find the answer without too much trouble.

The concept is fairly straightforward–replace unknown variables in a problem with certain values. When selecting a number, consider the following:

- Choose a number that's basic (just not too basic). Generally, you should avoid choosing 1 (or even 0). A decent choice is 2.

- Try not to choose a number that is displayed in the problem.

- Make sure you keep your numbers different if you need to choose at least two of them.

- If your question contains fractions, then a potential right answer may involve either an LCD (least common denominator) or an LCD multiple.

- 100 is the number you should choose when you are dealing with problems involving percentages.

ALEKS Math – Test Day Tips

After practicing and reviewing all the math concepts you've been taught, and taking some ALEKS mathematics practice tests, you'll be prepared for test day. Consider the following tips to be extra-ready come test time.

Before Your Test ....

What to do the night before:

- **Relax!** One day before your test, study lightly or skip studying altogether. You shouldn't attempt to learn something new, either. There are plenty of reasons why studying the evening before a big test can work against you. Put it this way–a marathoner wouldn't go out for a sprint before the day of a big race. Mental marathoners–such as yourself–should not study for any more than one hour 24 hours before a ALEKS test. That's because your brain requires some rest to be at its best. The night before your exam, spend some time with family or friends, or read a book.

- **Avoid bright screens** - You'll have to get some good shuteye the night before your test. Bright screens (such as the ones coming from your laptop, TV, or mobile device) should be avoided altogether. Staring at such a screen will keep your brain up, making it hard to drift asleep at a reasonable hour.

- **Make sure your dinner is healthy** - The meal that you have for dinner should be nutritious. Be sure to drink plenty of water as well. Load up on your complex carbohydrates, much like a marathon runner would do. Pasta, rice, and potatoes are ideal options here, as are vegetables and protein sources.

- **Get your bag ready for test day** – Prefer to take ALEKS in the Testing Office? The night prior to your test, pack your bag with your stationery, admissions pass, ID, and any other gear that you need. Keep the bag right by your front door. If you prefer to take the test at home, find a quite place without any distractions.

    Make plans to reach the testing site – If you are taking the test at the testing office, ensure that you understand precisely how you will arrive at the site of the test. If parking is something you'll have to find first, plan for it. If you're dependent on public transit, then review the schedule. You should also make sure that the train/bus/subway/streetcar you use will be running. Find out about road closures as well. If a parent or friend is accompanying you, ensure that they understand what steps they have to take as well.

The Day of the Test ....

- **Get up reasonably early, but not too early.**

- **Have breakfast** - Breakfast improves your concentration, memory, and mood. As such, make sure the breakfast that you eat in the morning is healthy. The last thing you want to be is distracted by a grumbling tummy. If it's not your own stomach making those noises, another test taker close to you might be instead. Prevent discomfort or embarrassment by consuming a healthy breakfast. Bring a snack with you if you think you'll need it.

- **Follow your daily routine** - Do you watch TV in the morning while getting ready for the day? Don't break your usual habits on the day of the test. Likewise, if coffee isn't something you drink in the morning, then don't take up the habit hours before your test. Routine consistency lets you concentrate on the main objective—doing the best you can on your test.

- **Wear layers** - Dress yourself up in comfortable layers if you are taking the test at the testing site. You should be ready for any kind of internal temperature. If it gets too warm during the test, take a layer off.

- **Make your voice heard** - If something is off, speak to a proctor. If medical attention is needed or if you'll require anything, consult the proctor prior to the start of the test. Any doubts you have should be clarified. You should be entering the test site with a state of mind that is completely clear.

- **Have faith in yourself** - When you feel confident, you will be able to perform at your best. When you are waiting for the test to begin, envision yourself receiving an outstanding result. Try to see yourself as someone who knows all the answers, no matter what the questions are. A lot of athletes tend to use this technique—particularly before a big competition. Your expectations will be reflected by your performance.

During your test

- **Be calm and breathe deeply** - You need to relax before the test, and some deep breathing will go a long way to help you do that. Be confident and calm. You got this. Everybody feels a little stressed out just before an evaluation of any kind is set to begin. Learn some effective breathing exercises. Spend a minute meditating before the test starts. Filter out any negative thoughts you have. Exhibit confidence when having such thoughts.

- **Concentrate on the test** - Refrain from comparing yourself to anyone else. You shouldn't be distracted by the people near you or random noise. Concentrate exclusively on the test. If you find yourself irritated by surrounding noises, earplugs can be used to block sounds off close to you. Don't forget–the test is going to last an hour or more. Some of that time will be dedicated to brief sections. Concentrate on the specific section you are working on during a particular moment. Do not let your mind wander off to upcoming or previous questions.

- **Try to answer each question individually** - Focus only on the question you are working on. Use one of the test-taking strategies to solve the problem. If you aren't able to come up with an answer, don't get frustrated. Simply guess, then move onto the next question.

- **Don't forget to breathe!** Whenever you notice your mind wandering, your stress levels boosting, or frustration brewing, take a thirty-second break. Shut your eyes, drop your pencil, breathe deeply, and let your shoulders relax. You will end up being more productive when you allow yourself to relax for a moment.

After your test

- **Take it easy** - You will need to set some time aside to relax and decompress once the test has concluded. There is no need to stress yourself out about what you could've said, or what you may have done wrong. At this point, there's nothing you can do about it. Your energy and time would be better spent on something that will bring you happiness for the remainder of your day.

- **Redoing the test** - Did you succeed on the test? Congratulations! Your hard work paid off! Succeeding on this test means that you are now ready to take college level courses.

  If you didn't receive the result you expected, though, don't worry! The test can be retaken. In such cases, you will need to follow the retake policy. You also need to re-register to take the exam again.

Contents

5 — Lines and Slope 51

6 — Polynomials 63

7 — Geometry, Solid Figures and Probability 77

1 Fundamentals and Building Blocks

Math topics that you'll learn in this chapter:

1. Adding and subtracting Integers
2. Multiplying and Dividing Integers
3. Order of Operations
4. Integers and Absolute Value
5. Percent Problems
6. Simple Interest

1

Adding and Subtracting Integers

☆ Integers include zero, counting numbers, and the negative of the counting numbers$\{..., -3, -2, -1, 0, 1, 2, 3, ...\}$

☆ Add a positive integer by moving to the right on the number line. (you will get a bigger number)

☆ Add a negative integer by moving to the left on the number line. (you will get a smaller number)

☆ Subtract an integer by adding its opposite.

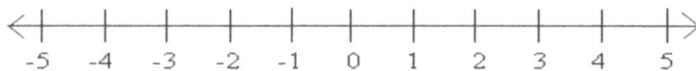

Examples:

Example 1. Solve. $(-3) - (-5) =$

Solution: Keep the first number and convert the sign of the second number to its opposite. (change subtraction into addition. Then: $(-3) + 5 = 2$

Example 1. Solve. $5 + (2 - 8) =$

Solution: First, subtract the numbers in brackets, $2 - 8 = -6$
Then: $5 + (-6) = \rightarrow$change addition into subtraction: $5 - 6 = -1$

Example 2. Solve. $(8 - 15) + 14 =$

Solution: First, subtract the numbers in brackets, $8 - 15 = -7$
Then: $-7 + 14 = \rightarrow -7 + 14 = 7$

Example 3. Solve. $18 + (-5 - 17) =$

Solution: First, subtract the numbers in brackets, $-5 - 17 = -22$
Then: $18 + (-22) = \rightarrow$change addition into subtraction: $18 - 22 = -4$

Multiplying and Dividing Integers

Use the following rules for multiplying and dividing integers:

☆ (negative) × (negative) = positive

☆ (negative) ÷ (negative) = positive

☆ (negative) × (positive) = negative

☆ (negative) ÷ (positive) = negative

☆ (positive) × (positive) = positive

☆ (positive) ÷ (negative) = negative

Examples:

Example 1. Solve. $5 \times (-2) =$

Solution: Use this rule: (positive) × (negative) = negative.
Then: $(5) \times (-2) = -10$

Example 2. Solve. $(-2) + (-30 \div 6) =$

Solution: First, divide -30 by 6, the numbers in brackets, use this rule:
(negative) ÷ (positive) = negative. Then: $-30 \div 6 = -5$
$(-2) + (-30 \div 6) = (-2) + (-5) = -2 - 5 = -7$

Example 3. Solve. $(13 - 16) \times (-3) =$

Solution: First, subtract the numbers in brackets,
$13 - 16 = -3 \rightarrow (-3) \times (-3) =$
Now use this rule: (negative) × (negative) = positive $\rightarrow (-3) \times (-3) = 9$

Example 4. Solve. $(18 - 3) \div (-5) =$

Solution: First, subtract the numbers in brackets,
$18 - 3 = 15 \rightarrow (15) \div (-5) =$
Now use this rule: (positive) ÷ (negative) = negative $\rightarrow (15) \div (-5) = -3$

Order of Operations

✫ In Mathematics, "operations" are addition, subtraction, multiplication, division, exponentiation (written as b^n) and grouping.

✫ When there is more than one math operation in an expression, use PEMDAS: (to memorize this rule, remember the phrase "Please Excuse My Dear Aunt Sally".)

- ❖ Parentheses
- ❖ Exponents
- ❖ Multiplication and Division (from left to right)
- ❖ Addition and Subtraction (from left to right)

Examples:

Example 1. Calculate. $(3 + 5) \div (8 \div 4) =$

Solution: First, simplify inside parentheses:
$(3 + 5) \div (8 \div 4) = (8) \div (8 \div 4) = (8) \div (2)$. Then: $(8) \div (2) = 4$

Example 2. Solve. $(4 \times 3) - (14 - 3) =$

Solution: First, calculate within parentheses: $(4 \times 3) - (14 - 3) = (12) - (11)$, Then: $(12) - (11) = 1$

Example 3. Calculate. $-3[(6 \times 5) \div (5 \times 3)] =$

Solution: First, calculate within parentheses:
$-3[(6 \times 5) \div (5 \times 3)] = -3[(30) \div (5 \times 3)] = -3[(30) \div (15)] = -3[2]$
Multiply -3 and 2. Then: $-3[2] = -6$

Example 4. Solve. $(32 \div 4) + (-25 + 5) =$

Solution: First, calculate within parentheses:
$(32 \div 4) + (-25 + 5) = (8) + (-20)$. Then: $(8) - (20) = -12$

Integers and Absolute Value

☆ The absolute value of a number is its distance from zero, in either direction, on the number line. For example, the distance of 9 and -9 from zero on number line is 9.

☆ The absolute value of an integer is the numerical value without its sign. (negative or positive)

☆ The vertical bar is used for absolute value as in $|x|$.

☆ The absolute value of a number is never negative; because it only shows, "how far the number is from zero".

Examples:

Example 1. Calculate. $|15 - 3| \times 6 =$

Solution: First, solve $|15 - 3| \to |15 - 3| = |12|$, the absolute value of 12 is 12, $|12| = 12$. Then: $12 \times 6 = 72$

Example 2. Solve. $|-35| \times |6 - 10| =$

Solution: First, find $|-35| \to$ the absolute value of -35 is 35. Then: $|-35| = 35$, $|-35| \times |6 - 10| =$
Now, calculate $|6 - 10| \to |6 - 10| = |-4|$, the absolute value of -4 is 4. $|-4| = 4$
Then: $35 \times 4 = 140$

Example 3. Solve. $|12 - 6| \times \frac{|-4 \times 5|}{3} =$

Solution: First, calculate $|12 - 6| \to |12 - 6| = |6|$, the absolute value of 6 is 6, $|6| = 6$. Then: $6 \times \frac{|-4 \times 5|}{3} =$
Now calculate $|-4 \times 5| \to |-4 \times 5| = |-20|$, the absolute value of -20 is 20, $|-20| = 20$. Then: $6 \times \frac{20}{3} = \frac{6 \times 20}{3} = \frac{120}{3} = 40$

Percent Problems

☆ Percent is a ratio of a number and 100. It always has the same denominator, 100. The percent symbol is "%".

☆ Percent means "per 100". So, 20% is $\frac{20}{100}$.

☆ In each percent problem, we are looking for the base, or the part or the percent.

☆ Use these equations to find each missing section in a percent problem:

    ❖ Base = Part ÷ Percent

    ❖ Part = Percent × Base
    ❖ Percent = Part ÷ Base

Examples:

Example 1. What is 30% of 60?

Solution: In this problem, we have the percent (30%) and the base (60) and we are looking for the "part". Use this formula: *Part = Percent × Base*.
Then: $Part = 30\% \times 60 = \frac{30}{100} \times 60 = 0.30 \times 60 = 18$. The answer: 30% of 60 is 18.

Example 2. 20 is what percent of 400?

Solution: In this problem, we are looking for the percent. Use this equation:
$Percent = Part \div Base \rightarrow Percent = 20 \div 400 = 0.05 = 5\%$.
Then: 20 is 5 percent of 400.

Example 3. 70 is 25 percent of what number?

Solution: In this problem, we are looking for the base. Use this equation:
$Base = Part \div Percent \rightarrow Base = 70 \div 25\% = 70 \div 0.25 = 280$
Then: 70 is 25 percent of 280.

Simple Interest

☆ Simple Interest: The charge for borrowing money or the return for lending it.

☆ Simple interest is calculated on the initial amount (principal).

☆ To solve a simple interest problem, use this formula:

$$Interest = principal \times rate \times time \rightarrow (I = p \times r \times t = prt)$$

Examples:

Example 1. Find simple interest for $250 investment at 6% for 5 years.

Solution: Use Interest formula:
$I = prt$ ($P = \$250, r = 6\% = \frac{6}{100} = 0.06$ and $t = 5$)
Then: $I = 250 \times 0.06 \times 5 = \$75$

Example 2. Find simple interest for $1,300 at 3% for 2 years.

Solution: Use Interest formula:
$I = prt$ ($P = \$1,300, r = 3\% = \frac{3}{100} = 0.03$ and $t = 2$)
Then: $I = 1,300 \times 0.03 \times 2 = \$78.00$

Example 3. Andy received a student loan to pay for his educational expenses this year. What is the interest on the loan if he borrowed $5,100 at 5% for 4 years?

Solution: Use Interest formula: $I = prt$. $P = \$5,100, r = 5\% = 0.05$ and $t = 4$
Then: $I = 5,100 \times 0.05 \times 4 = \$1,020$

Example 4. Bob is starting his own small business. He borrowed $28,000 from the bank at an 7% rate for 6 months. Find the interest Bob will pay on this loan.

Solution: Use Interest formula:
$I = prt$. $P = \$28,000, r = 7\% = 0.07$ and $t = 0.5$ (6 months is half year). Then:
$I = 28,000 \times 0.07 \times 0.5 = \$980$

Day 1: Practices

✎ **Find each sum or difference.**

1) $-6 + 17 =$

2) $12 - 21 =$

3) $31 - (-4) =$

4) $(7 + 5) + (8 - 3) =$

5) $(2 - 3) - (15 - 11) =$

6) $(-8 - 7) - (-6 - 2) =$

✎ **Solve.**

7) $2 \times (-4) =$

8) $(-5) \times (-3) =$

9) $(-15) \div 5 =$

10) $(-4) \times (-5) \times (-2) =$

11) $(-6 + 36) \div (-2) =$

12) $(-25 + 5) \times (-5 - 3) =$

✎ **Evaluate each expression.**

13) $5 - (2 \times 3) =$

14) $(6 \times 5) - 8 =$

15) $(-5 \times 3) + 4 =$

16) $(-35 \div 5) - (12 + 2) =$

17) $4 \times [(2 \times 3) \div (-3 + 1)] =$

18) $35 \div [(6 - 1) \times (7 - 8)] =$

✎ **Find the answers.**

19) $|-3| + |7 - 9| =$

20) $|8 - 10| + |6 - 7| =$

21) $|-6 + 10| - |-9 - 3| =$

22) $3 + |2 - 1| + |3 - 12| =$

23) $-8 - |3 - 6| + |2 + 3| =$

24) $|-6| \times |-5.4| =$

25) $|3 \times (-4)| \times \frac{8}{3} =$

26) $|(-2) \times (-2)| \times \frac{1}{4} =$

27) $|-8| + |(-8) \times 2| =$

28) $|(-3) \times (-5)| \times |(-3) \times (-4)| =$

✎ Find the answers.

29) Round 4.2873 to the thousandth-place value

30) $[6 \times (-16) + 8] - (-4) + [4 \times 5] \div 2 =$

✎ Solve each problem.

31) What is 15% of 60? ____

32) What is 20% of 500? ____

33) 25 what is percent of 250? ___

34) 30 is what percent of 150? ___

35) 15 is 10 percent of what number? ___

36) 25 is 5 percent of what number? ___

✎ Determine the simple interest for the following loans.

37) $800 at 3% for 2 years. $___

38) $260 at 10% for 5 years. $___

39) $380 at 4% for 3 years. $___

40) $1,200 at 2% for 1 years. $___

Day 1: Answers

1) $-6 + 17 = 17 - 6 = 11$

2) $12 - 21 = -9$

3) $31 - (-4) = 31 + 4 = 35$

4) $(7 + 5) + (8 - 3) = 12 + 5 = 17$

5) $(2 - 3) - (15 - 11) = (-1) - (4) = -1 - 4 = -5$

6) $(-8 - 7) - (-6 - 2) = (-15) - (-8) = -15 + 8 = -7$

7) Use this rule: (positive) × (negative) = negative → $2 \times (-4) = -8$

8) Use this rule: (negative) × (negative) = positive → $(-5) \times (-3) = 15$

9) Use this rule: (negative) ÷ (positive) = negative → $(-15) \div 5 = -3$

10) Use these rules: [(negative) × (negative) = positive] and [(positive) × (negative) = negative]→ $(-4) \times (-5) \times (-2) = (20) \times (-2) = -40$

11) Use this rule: (positive)÷ (negative) = negative→
$(-6 + 36) \div (-2) = (30) \div (-2) = -15$

12) Use this rule: (negative) × (negative) = positive→
$(-25 + 5) \times (-5 - 3) = (-20) \times (-8) = -160$

13) $5 - (2 \times 3) = 5 - 6 = -1$

14) $(6 \times 5) - 8 = 30 - 8 = 22$

15) Use this rule: (negative) × (positive) = negative → $(-5 \times 3) + 4 = -15 + 4 = -11$

16) Use this rule: (negative) ÷ (positive) = negative →
$(-35 \div 5) - (12 + 2) = -7 - 14 = -21$

17) Use these rules: [(positive) × (negative) = negative] and [(positive) ÷ (negative) = negative]→ $4 \times [(2 \times 3) \div (-3 + 1)] = 4 \times [6 \div (-2)] = 4 \times (-3) = -12$

18) Use these rules: [(positive) × (negative) = negative] and [(positive) ÷ (negative) = negative]→ $35 \div [(6 - 1) \times (7 - 8)] = 35 \div [5 \times (-1)] = 35 \div (-5) = -7$

19) $|-3| + |7 - 9| = 3 + |-2| = 3 + 2 = 5$

20) $|8 - 10| + |6 - 7| = |-2| + |-1| = 2 + 1 = 3$

21) $|-6 + 10| - |-9 - 3| = |4| - |-12| = 4 - (12) = 4 - 12 = -8$

22) $3 + |2 - 1| + |3 - 12| = 3 + |1| + |-9| = 3 + 1 + (9) = 3 + 1 + 9 = 13$

23) $-8 - |3 - 6| + |2 + 3| = -8 - |-3| + |5| = -8 - (3) + 5 = -8 - 3 + 5 = -11 + 5 = -6$

24) $|-6| \times |-5.4| = 6 \times 5.4 = 32.4$

25) $|3 \times (-4)| \times \frac{8}{3} = |-12| \times \frac{8}{3} = 12 \times \frac{8}{3} = \frac{12 \times 8}{3} = 32$

26) $|(-2) \times (-2)| \times \frac{1}{4} = |4| \times \frac{1}{4} = 4 \times \frac{1}{4} = 1$

27) $|-8| + |(-8) \times 2| = 8 + |-16| = 8 + 16 = 24$

28) $|(-3) \times (-5)| \times |(-3) \times (-4)| = |15| \times |12| = 15 \times 12 = 180$

29) $4.2873 \approx 4.287$

30) $[6 \times (-16) + 8] - (-4) + [4 \times 5] \div 2 = [(-96) + 8] - (-4) + [4 \times 5] \div 2 = (-88) - (-4) + (20) \div 2 = (-88) - (-4) + 10 = (-88) + 4 + (10) = -84 + 10 = -74$

31) Part = Percent $\times$ Base $\rightarrow 15\% \times 60 = \frac{15}{100} \times 60 = 0.15 \times 60 = 9$

32) Part = Percent $\times$ Base $\rightarrow 20\% \times 500 = \frac{20}{100} \times 500 = 0.2 \times 500 = 100$

33) Percent = Part $\div$ Base $\rightarrow 25 \div 250 = \frac{25}{250} = \frac{25 \div 25}{250 \div 25} = \frac{1}{10} = \frac{1}{10} \times 100 = 10\%$

34) Percent = Part $\div$ Base $\rightarrow 30 \div 150 = \frac{30}{150} = \frac{30 \div 30}{150 \div 30} = \frac{1}{5} = \frac{1}{5} \times 100 = 20\%$

35) Base = Part $\div$ Percent $\rightarrow 15 \div 10\% = 15 \div \frac{10}{100} = 15 \div 0.1 = 150$

36) Base = Part $\div$ Percent $\rightarrow 25 \div 5\% = 25 \div \frac{5}{100} = 25 \div 0.05 = 500$

37) $800 \times 3\% \times 2 = 800 \times \frac{3}{100} \times 2 = \frac{4,800}{100} = 48$

38) $260 \times 10\% \times 5 = 260 \times \frac{10}{100} \times 5 = 260 \times \frac{1}{10} \times 5 = 130$

39) $380 \times 4\% \times 3 = 380 \times \frac{4}{100} \times 3 = 45.6$

40) $1,200 \times 2\% \times 1 = 1,200 \times \frac{2}{100} \times 1 = 24$

2 Exponents and Variables

Math topics that you'll learn in this chapter:

13

Multiplication Property of Exponents

☆ Exponents are shorthand for repeated multiplication of the same number by itself. For example, instead of 2×2, we can write 2^2. For $3 \times 3 \times 3 \times 3$, we can write 3^4.

☆ In algebra, a variable is a letter used to stand for a number. The most common letters are: $x, y, z, a, b, c, m,$ and n.

☆ Exponent's rules: $x^a \times x^b = x^{a+b}$, $\dfrac{x^a}{x^b} = x^{a-b}$

$$(x^a)^b = x^{a \times b} \qquad\qquad (xy)^a = x^a \times y^a \qquad\qquad \left(\dfrac{a}{b}\right)^c = \dfrac{a^c}{b^c}$$

Examples:

Example 1. *Multiply.* $3x^2 \times 4x^3$

Solution: *Use Exponent's rules*: $x^a \times x^b = x^{a+b} \rightarrow x^2 \times x^3 = x^{2+3} = x^5$
Then: $3x^2 \times 4x^3 = 12x^5$

Example 2. *Simplify.* $(x^2y^4)^3$

Solution: *Use Exponent's rules*: $(x^a)^b = x^{a \times b}$.
Then: $(x^2y^4)^3 = x^{2 \times 3}y^{4 \times 3} = x^6y^{12}$

.

Example 3. *Multiply.* $6x^7 \times 4x^3$

Solution: *Use Exponent's rules*: $x^a \times x^b = x^{a+b} \rightarrow x^7 \times x^3 = x^{7+3} = x^{10}$
Then: $6x^7 \times 4x^3 = 24x^{10}$

Example 4. *Simplify.* $\left(x^2y^5\right)^4$

Solution: *Use Exponent's rules*: $(x^a)^b = x^{a \times b}$.
Then: $\left(x^2y^5\right)^4 = x^{2 \times 4}y^{5 \times 4} = x^8y^{20}$

Division Property of Exponents

For division of exponents use following formulas:

☆ $\frac{x^a}{x^b} = x^{a-b}$ $(x \neq 0)$

☆ $\frac{x^a}{x^b} = \frac{1}{x^{b-a}}$, $(x \neq 0)$

☆ $\frac{1}{x^b} = x^{-b}$

Examples:

Example 1. Simplify. $\frac{12x^2y}{3xy^3} =$

Solution: First, cancel the common factor: $3 \rightarrow \frac{12x^2y}{3xy^3} = \frac{4x^2y}{xy^3}$

Use Exponent's rules: $\frac{x^a}{x^b} = x^{a-b} \rightarrow \frac{x^2}{x} = x^{2-1} = x^1$ and $\frac{x^a}{x^b} = \frac{1}{x^{b-a}} \rightarrow \frac{y}{y^3} = \frac{1}{y^{3-1}} = \frac{1}{y^2}$

Then: $\frac{12x^2y}{3xy^3} = \frac{4x}{y^2}$

Example 2. Simplify. $\frac{48x^{12}}{16x^9} =$

Solution: Use Exponent's rules: $\frac{x^a}{x^b} = x^{a-b} \rightarrow \frac{x^{12}}{x^9} = x^{12-9} = x^3$

Then: $\frac{48x^{12}}{16x^9} = 3x^3$

Example 3. Simplify. $\frac{6x^5y^7}{42x^8y^2} =$

Solution: First, cancel the common factor: $6 \rightarrow \frac{x^5y^7}{7x^8y^2}$

Use Exponent's rules: $\frac{x^a}{x^b} = x^{a-b} \rightarrow \frac{x^5}{x^8} = x^{5-8} = x^{-3} = \frac{1}{x^3}$ and $\frac{y^7}{y^2} = y^{7-2} = y^5$

Then: $\frac{6x^5y^7}{42x^8y^2} = \frac{y^5}{7x^3}$

Powers of Products and Quotients

☆ For any nonzero numbers a and b and any integer x, $(ab)^x = a^x \times b^x$

and $\left(\frac{a}{b}\right)^c = \frac{a^c}{b^c}$

Examples:

Example 1. Simplify. $\left(6x^4y^5\right)^2$

Solution: Use Exponent's rules: $(x^a)^b = x^{a \times b}$

$\left(6x^4y^5\right)^2 = (6)^2(x^4)^2\left(y^5\right)^2 = 36x^{4\times2}y^{5\times2} = 36x^8y^{10}$

Example 2. Simplify. $\left(\frac{5x^5}{2x^4}\right)^2$

Solution: First, cancel the common factor: $x^4 \rightarrow \left(\frac{5x^5}{2x^4}\right) = \left(\frac{5x}{2}\right)^2$

Use Exponent's rules: $\left(\frac{a}{b}\right)^c = \frac{a^c}{b^c}$. Then: $\left(\frac{5x}{2}\right)^2 = \frac{(5x)^2}{(2)^2} = \frac{25x^2}{4}$

Example 3. Simplify. $(-6x^7y^3)^2$

Solution: Use Exponent's rules: $(x^a)^b = x^{a \times b}$

$(-6x^7y^3)^2 = (-6)^2(x^7)^2(y^3)^2 = 36x^{7\times2}y^{3\times2} = 36x^{14}y^6$

Example 4. Simplify. $\left(\frac{8x^3}{5x^7}\right)^2$

Solution: First, cancel the common factor: $x^3 \rightarrow \left(\frac{8x^3}{5x^7}\right)^2 = \left(\frac{8}{5x^4}\right)^2$

Use Exponent's rules: $\left(\frac{a}{b}\right)^c = \frac{a^c}{b^c}$, Then: $\left(\frac{8}{5x^4}\right)^2 = \frac{8^2}{\left(5x^4\right)^2} = \frac{64}{25x^8}$

Zero and Negative Exponents

☆ Zero-Exponent Rule: $a^0 = 1$, this means that anything raised to the zero power is 1. For example: $(5xy)^0 = 1$ (number zero is an exception: $0^0 = 0$)

☆ A negative exponent simply means that the base is on the wrong side of the fraction line, so you need to flip the base to the other side. For instance, "x^{-2}" (pronounced as "ecks to the minus two") just means "x^2" but underneath, as in $\frac{1}{x^2}$.

Examples:

Example 1. *Evaluate.* $\left(\frac{3}{7}\right)^{-2} =$

Solution: Use negative exponent's rule: $\left(\frac{x^a}{x^b}\right)^{-2} = \left(\frac{x^b}{x^a}\right)^2 \rightarrow \left(\frac{3}{7}\right)^{-2} = \left(\frac{7}{3}\right)^2$
Then: $\left(\frac{7}{3}\right)^2 = \frac{7^2}{3^2} = \frac{49}{9}$

Example 2. *Evaluate.* $\left(\frac{2}{5}\right)^{-3} =$

Solution: Use negative exponent's rule: $\left(\frac{x^a}{x^b}\right)^{-3} = \left(\frac{x^b}{x^a}\right)^3 \rightarrow \left(\frac{2}{5}\right)^{-3} = \left(\frac{5}{2}\right)^3 =$
Then: $\left(\frac{5}{2}\right)^3 = \frac{5^3}{2^3} = \frac{125}{8}$

Example 3. *Evaluate.* $\left(\frac{a}{b}\right)^0 =$

Solution: Use zero-exponent Rule: $a^0 = 1$
Then: $\left(\frac{a}{b}\right)^0 = 1$

Example 4. *Evaluate.* $\left(\frac{9}{5}\right)^{-1} =$

Solution: Use negative exponent's rule: $\left(\frac{x^a}{x^b}\right)^{-1} = \left(\frac{x^b}{x^a}\right)^1 \rightarrow \left(\frac{9}{5}\right)^{-1} = \left(\frac{5}{9}\right)^1 = \frac{5}{9}$

Negative Exponents and Negative Bases

☆ A negative exponent is the reciprocal of that number with a positive exponent. $(3)^{-2} = \frac{1}{3^2}$

☆ To simplify a negative exponent, make the power positive!

☆ The parenthesis is important! -5^{-2} is not the same as $(-5)^{-2}$

$$-5^{-2} = -\frac{1}{5^2} \text{ and } (-5)^{-2} = +\frac{1}{5^2}$$

Examples:

Example 1. *Simplify.* $\left(\frac{4a}{7c}\right)^{-2} =$

Solution: Use negative exponent's rule: $\left(\frac{x^a}{x^b}\right)^{-2} = \left(\frac{x^b}{x^a}\right)^{2} \rightarrow \left(\frac{4a}{7c}\right)^{-2} = \left(\frac{7c}{4a}\right)^{2}$

Now use exponent's rule: $\left(\frac{a}{b}\right)^c = \frac{a^c}{b^c} \rightarrow \left(\frac{7c}{4a}\right)^{2} = \frac{7^2 c^2}{4^2 a^2}$

Then: $\frac{7^2 c^2}{4^2 a^2} = \frac{49 c^2}{16 a^2}$

Example 2. *Simplify.* $\left(\frac{3x}{y}\right)^{-3} =$

Solution: Use negative exponent's rule: $\left(\frac{x^a}{x^b}\right)^{-3} = \left(\frac{x^b}{x^a}\right)^{3} \rightarrow \left(\frac{3x}{y}\right)^{-3} = \left(\frac{y}{3x}\right)^{3}$

Now use exponent's rule: $\left(\frac{a}{b}\right)^c = \frac{a^c}{b^c} \rightarrow \left(\frac{y}{3x}\right)^{3} = \frac{y^3}{3^3 x^3} = \frac{y^3}{27 x^3}$

Example 3. *Simplify.* $\left(\frac{7a}{4c}\right)^{-2} =$

Solution: Use negative exponent's rule: $\left(\frac{x^a}{x^b}\right)^{-2} = \left(\frac{x^b}{x^a}\right)^{2} \rightarrow \left(\frac{7a}{4c}\right)^{-2} = \left(\frac{4c}{7a}\right)^{2}$

Now use exponent's rule: $\left(\frac{a}{b}\right)^c = \frac{a^c}{b^c} \rightarrow \left(\frac{4c}{7a}\right)^{2} = \frac{4^2 c^2}{7^2 a^2}$

Then: $\frac{4^2 c^2}{7^2 a^2} = \frac{16 c^2}{49 a^2}$

www.EffortlessMath.com

Scientific Notation

☆ Scientific notation is used to write very big or very small numbers in decimal form.

☆ In scientific notation, all numbers are written in the form of: $m \times 10^n$, where m is greater than 1 and less than 10.

☆ To convert a number from scientific notation to standard form, move the decimal point to the left (if the exponent of ten is a negative number), or to the right (if the exponent is positive).

Examples:

Example 1. *Write* 0.000037 *in scientific notation.*

Solution: First, move the decimal point to the right so you have a number between 1 and 10. That number is 3.7. Now, determine how many places the decimal moved in step 1 by the power of 10. We moved the decimal point 5 digits to the right. Then: 10^{-5}. When the decimal moved to the right, the exponent is negative. Then: $0.000037 = 3.7 \times 10^{-5}$

Example 2. *Write* 5.3×10^{-3} *in standard notation.*

Solution: The exponent is negative 3. Then, move the decimal point to the left three digits. (remember 5.3 = 0000005.3) When the decimal moved to the right, the exponent is negative. Then: $5.3 \times 10^{-3} = 0.0053$

Example 3. *Write* 0.00042 *in scientific notation.*

Solution: First, move the decimal point to the right so you have a number between 1 and 10. Then: $m = 4.2$. Now, determine how many places the decimal moved in step 1 by the power of 10. 10^{-4}. Then: $0.00042 = 4.2 \times 10^{-4}$

Example 4. *Write* 7.3×10^7 *in standard notation.*

Solution: The exponent is positive 7. Then, move the decimal point to the right seven digits. (remember 7.3 = 7.3000000) Then: $7.3 \times 10^7 = 73,000,000$

Radicals

☆ If n is a positive integer and x is a real number, then: $\sqrt[n]{x} = x^{\frac{1}{n}}$,

$$\sqrt[n]{xy} = x^{\frac{1}{n}} \times y^{\frac{1}{n}}, \sqrt[n]{\frac{x}{y}} = \frac{x^{\frac{1}{n}}}{y^{\frac{1}{n}}}, \text{ and } \sqrt[n]{x} \times \sqrt[n]{y} = \sqrt[n]{xy}$$

☆ A square root of x is a number r whose square is: $r^2 = x$ (r is a square root of x)

☆ To add and subtract radicals, we need to have the same values under the radical. For example: $\sqrt{5} + 3\sqrt{5} = 4\sqrt{5}$, $5\sqrt{6} - 2\sqrt{6} = 3\sqrt{5}$

Examples:

Example 1. Find the square root of $\sqrt{256}$.

Solution: First, factor the number: $256 = 16^2$. Then: $\sqrt{256} = \sqrt{16^2}$
Now use radical rule: $\sqrt[n]{a^n} = a$. Then: $\sqrt{256} = \sqrt{16^2} = 16$

Example 2. Evaluate. $\sqrt{9} \times \sqrt{36} =$

Solution: Find the values of $\sqrt{9}$ and $\sqrt{36}$. Then: $\sqrt{9} \times \sqrt{36} = 3 \times 6 = 18$

Example 3. Solve. $3\sqrt{7} + 11\sqrt{7}$.

Solution: Since we have the same values under the radical, we can add these two radicals: $3\sqrt{7} + 11\sqrt{7} = 14\sqrt{7}$

Example 4. Evaluate. $\sqrt{40} \times \sqrt{10} =$

Solution: Use this radical rule: $\sqrt[n]{x} \times \sqrt[n]{y} = \sqrt[n]{xy} \rightarrow \sqrt{40} \times \sqrt{10} = \sqrt{400}$
The square root of 400 is 20. Then: $\sqrt{40} \times \sqrt{10} = \sqrt{400} = 20$

Day 2: Practices

✍ Find the products.

1) $5x^3 \times 2x =$

2) $x^4 \times 5x^2y =$

3) $2xy \times 3x^5y^2 =$

4) $4xy^2 \times 2x^2y =$

5) $-3x^3y^3 \times 2x^2y^2 =$

6) $-5xy^2 \times 3x^5y^2 =$

7) $-5x^2y^6 \times 6x^5y^2 =$

8) $-2x^3y^3 \times 2x^3y^3 =$

9) $-7xy^3 \times 4x^5y^2 =$

10) $-x^4y^3 \times (-5x^6y^2) =$

11) $-6y^6 \times 7x^6y^2 =$

12) $-8x^4 \times 2y^2 =$

✍ Simplify.

13) $\frac{3^2 \times 3^3}{3^3 \times 3} =$

14) $\frac{4^2 \times 4^4}{5^4 \times 5} =$

15) $\frac{14x^5}{7x^2} =$

16) $\frac{15x^3}{5x^6} =$

17) $\frac{64y^3}{8xy^7} =$

18) $\frac{10x^4y^5}{30x^5y^4} =$

19) $\frac{11y}{44x^3y^3} =$

20) $\frac{40xy^3}{120xy^3} =$

21) $\frac{45x^3}{25xy^3} =$

22) $\frac{72y^6x}{36x^8y^9} =$

✍ Solve.

23) $(x^2 \, y^2)^3 =$

24) $(2x^3 \, y^2)^3 =$

25) $(2x \times 3xy^2)^2 =$

26) $(4x \times 2y^4)^2 =$

27) $\left(\frac{3x}{x^2}\right)^2 =$

28) $\left(\frac{6y}{18y^3}\right)^2 =$

29) $\left(\frac{3x^2y^2}{12x^4y^3}\right)^3 =$

30) $\left(\frac{23x^5y^3}{46x^3y^5}\right)^3 =$

31) $\left(\frac{16x^7y^3}{48x^5y^2}\right)^2 =$

32) $\left(\frac{12x^5y^6}{60x^7y^2}\right)^2 =$

✎ Evaluate each expression. (Zero and Negative Exponents)

33) $\left(\frac{1}{3}\right)^{-2} =$

34) $\left(\frac{1}{4}\right)^{-3} =$

35) $\left(\frac{1}{6}\right)^{-2} =$

36) $\left(\frac{2}{3}\right)^{-3} =$

37) $\left(\frac{2}{5}\right)^{-3} =$

38) $\left(\frac{3}{5}\right)^{-2} =$

✎ Write each expression with positive exponents.

39) $2y^{-3} =$

40) $13y^{-5} =$

41) $-20x^{-2} =$

42) $15a^{-2}b^3 =$

43) $23a^2b^{-4}c^{-8} =$

44) $-4x^4y^{-2} =$

45) $\frac{16y}{x^3y^{-4}} =$

46) $\frac{30a^{-3}b}{-100c^{-2}} =$

✎ Write each number in scientific notation.

47) $0.00518 =$

48) $0.000042 =$

49) $78,000 =$

50) $92,000,000 =$

✎ Evaluate.

51) $\sqrt{5} \times \sqrt{5} =$

52) $\sqrt{25} - \sqrt{4} =$

53) $\sqrt{81} + \sqrt{36} =$

54) $\sqrt{4} \times \sqrt{25} =$

55) $\sqrt{2} \times \sqrt{18} =$

56) $4\sqrt{2} + 3\sqrt{2} =$

57) $5\sqrt{7} + 2\sqrt{7} =$

58) $\sqrt{45} + 2\sqrt{5} =$

Day 2: Answers

1) $5x^3 \times 2x \rightarrow x^3 \times x^1 = x^{3+1} = x^4 \rightarrow 5x^3 \times 2x = 10x^4$

2) $x^4 \times 5x^2y \rightarrow x^4 \times x^2 = x^{4+2} = x^6 \rightarrow x^4 \times 5x^2y = 5x^6y$

3) $2xy \times 3x^5y^2 \rightarrow x \times x^5 = x^{1+5} = x^6,\ y \times y^2 = y^{1+2} = y^3 \rightarrow 2xy \times 3x^5y^2 = 6x^6y^3$

4) $4xy^2 \times 2x^2y \rightarrow x \times x^2 = x^{1+2} = x^3,\ y^2 \times y = y^{2+1} = y^3 \rightarrow 4xy^2 \times 2x^2y = 8x^3y^3$

5) $-3x^3y^3 \times 2x^2y^2 \rightarrow x^3 \times x^2 = x^{3+2} = x^5,\ y^3 \times y^2 = y^{3+2} = y^5 \rightarrow$

   $-3x^3y^3 \times 2x^2y^2 = -6x^5y^5$

6) $-5xy^2 \times 3x^5y^2 \rightarrow x \times x^5 = x^{1+5} = x^6,\ y^2 \times y^2 = y^{2+2} = y^4 \rightarrow$

   $-5xy^2 \times 3x^5y^2 = -15x^6y^4$

7) $-5x^2y^6 \times 6x^5y^2 \rightarrow x^2 \times x^5 = x^{2+5} = x^7,\ y^6 \times y^2 = y^{6+2} = y^8 \rightarrow$

   $-5x^2y^6 \times 6x^5y^2 = -30x^7y^8$

8) $-2x^3y^3 \times 2x^3y^3 \rightarrow x^3 \times x^3 = x^{3+3} = x^6,\ y^3 \times y^3 = y^{3+3} = y^6 \rightarrow$

   $-2x^3y^3 \times 2x^3y^3 = -4x^6y^6$

9) $-7xy^3 \times 4x^5y^2 \rightarrow x \times x^5 = x^{1+5} = x^6,\ y^3 \times y^2 = y^{3+2} = y^5 \rightarrow$

   $-7xy^3 \times 4x^5y^2 = -28x^6y^5$

10) $-x^4y^3 \times (-5x^6y^2) \rightarrow x^4 \times x^6 = x^{4+6} = x^{10},\ y^3 \times y^2 = y^{3+2} = y^5 \rightarrow$

   $-x^4y^3 \times (-5x^6y^2) = 5x^{10}y^5$

11) $-6y^6 \times 7x^6y^2 \rightarrow y^6 \times y^2 = y^{6+2} = y^8 \rightarrow -6y^6 \times 7x^6y^2 = -42x^6y^8$

12) $-8x^4 \times 2y^2 = -16x^4y^2$

13) $\frac{3^2 \times 3^3}{3^3 \times 3} = \frac{3^{2+3}}{3^{3+1}} = \frac{3^5}{3^4} = 3^{5-4} = 3^1 = 3$

14) $\frac{4^2 \times 4^4}{5^4 \times 5} = \frac{4^{2+4}}{5^{4+1}} = \frac{4^6}{5^5}$

15) $\frac{14x^5}{7x^2} \rightarrow \frac{14 \div 7}{7 \div 7} = 2,\ \frac{x^5}{x^2} = x^{5-2} = x^3 \rightarrow \frac{14x^5}{7x^2} = 2x^3$

16) $\frac{15x^3}{5x^6} \rightarrow \frac{15 \div 5}{5 \div 5} = 3,\ \frac{x^3}{x^6} = x^{3-6} = x^{-3} = \frac{1}{x^3} \rightarrow \frac{15x^3}{5x^6} = \frac{3}{x^3}$

17) $\frac{64y^3}{8xy^7} \rightarrow \frac{64 \div 8}{8 \div 8} = 8,\ \frac{y^3}{y^7} = y^{3-7} = y^{-4} = \frac{1}{y^4} \rightarrow \frac{64y^3}{8xy^7} = \frac{8}{xy^4}$

18) $\frac{10x^4y^5}{30x^5y^4} \rightarrow \frac{10 \div 10}{30 \div 10} = \frac{1}{3},\ \frac{x^4}{x^5} = x^{4-5} = x^{-1} = \frac{1}{x},\ \frac{y^5}{y^4} = y^{5-4} = y^1 \rightarrow \frac{10x^4y^5}{30x^5y^4} = \frac{y}{3x}$

19) $\frac{11y}{44x^3y^3} \rightarrow \frac{11 \div 11}{44 \div 11} = \frac{1}{4}, \frac{y}{y^3} = y^{1-3} = y^{-2} = \frac{1}{y^2} \rightarrow \frac{11y}{44x^3y^3} = \frac{1}{4x^3y^2}$

20) $\frac{40xy^3}{120xy^3} \rightarrow \frac{40 \div 40}{120 \div 40} = \frac{1}{3}, \frac{x}{x} = x^{1-1} = x^0 = 1, \frac{y^3}{y^3} = y^{3-3} = y^0 = 1 \rightarrow \frac{40xy^3}{120xy^3} = \frac{1}{3}$

21) $\frac{45x^3}{25xy^3} \rightarrow \frac{45 \div 5}{25 \div 5} = \frac{9}{5}, \frac{x^3}{x} = x^{3-1} = x^2 \rightarrow \frac{45x^3}{25xy^3} = \frac{9x^2}{5y^3}$

22) $\frac{72y^6x}{36x^8y^9} \rightarrow \frac{72 \div 36}{36 \div 36} = \frac{2}{1} = 2, \frac{y^6}{y^9} = y^{6-9} = y^{-3} = \frac{1}{y^3}, \frac{x}{x^8} = x^{1-8} = x^{-7} = \frac{1}{x^7} \rightarrow \frac{72y^6x}{36x^8y^9} = \frac{2}{x^7y^3}$

23) $(x^2 y^2)^3 = x^{2 \times 3} y^{2 \times 3} = x^6 y^6$

24) $(2x^3 y^2)^3 = 2^3 x^{3 \times 3} y^{2 \times 3} = 8x^9 y^6$

25) $(2x \times 3xy^2)^2 = (2 \times 3)^2 x^{(1+1) \times 2} y^{2 \times 2} = 6^2 x^{2 \times 2} y^4 = 36x^4 y^4$

26) $(4x \times 2y^4)^2 = (4 \times 2)^2 x^{1 \times 2} y^{4 \times 2} = 8^2 x^2 y^8 = 64x^2 y^8$

27) $\left(\frac{3x}{x^2}\right)^2 = \frac{3^{1 \times 2} x^{1 \times 2}}{x^{2 \times 2}} = \frac{3^2 x^2}{x^4} \rightarrow 3^2 = 9, \frac{x^2}{x^4} = x^{2-4} = x^{-2} = \frac{1}{x^2} \rightarrow \left(\frac{3x}{x^2}\right)^2 = \frac{9}{x^2}$

28) $\left(\frac{6y}{18y^3}\right)^2 = \frac{6^{1 \times 2} y^{1 \times 2}}{(18)^{1 \times 2} y^{3 \times 2}} = \frac{6^2 y^2}{18^2 y^6} \rightarrow \frac{36}{324} = \frac{1}{9}, \frac{y^2}{y^6} = y^{2-6} = y^{-4} = \frac{1}{y^4} \rightarrow \left(\frac{6y}{18y^3}\right)^2 = \frac{1}{9y^4}$

29) $\left(\frac{3x^2y^2}{12x^4y^3}\right)^3 = \frac{3^{1 \times 3} x^{2 \times 3} y^{2 \times 3}}{(12)^{1 \times 3} x^{4 \times 3} y^{3 \times 3}} = \frac{3^3 x^6 y^6}{12^3 x^{12} y^9} \rightarrow \frac{3^3}{12^3} = \frac{27}{1,728} = \frac{1}{64}, \frac{x^6}{x^{12}} = x^{6-12} = x^{-6} = \frac{1}{x^6}, \frac{y^6}{y^9} =$

$y^{6-9} = y^{-3} = \frac{1}{y^3} \rightarrow \left(\frac{3x^2y^2}{12x^4y^3}\right)^3 = \frac{1}{64x^6y^3}$

30) $\left(\frac{23x^5y^3}{46x^3y^5}\right)^3 = \frac{23^{1 \times 3} x^{5 \times 3} y^{3 \times 3}}{(23 \times 2)^{1 \times 3} x^{3 \times 3} y^{5 \times 3}} = \frac{23^3 x^{15} y^9}{23^3 \times 2^3 x^9 y^{15}} = \frac{x^{15} y^9}{8x^9 y^{15}} \rightarrow \frac{x^{15}}{x^9} = x^{15-9} = x^6, \frac{y^9}{y^{15}} = y^{9-15} =$

$y^{-6} = \frac{1}{y^6} \rightarrow \left(\frac{23x^5y^3}{46x^3y^5}\right)^3 = \frac{x^6}{8y^6}$

31) $\left(\frac{16x^7y^3}{48x^5y^2}\right)^2 = \frac{16^{1 \times 2} x^{7 \times 2} y^{3 \times 2}}{(16 \times 3)^{1 \times 2} x^{5 \times 2} y^{2 \times 2}} = \frac{16^2 x^{14} y^6}{16^2 \times 3^2 x^{10} y^4} = \frac{x^{14} y^6}{9x^{10} y^4} \rightarrow \frac{x^{14}}{x^{10}} = x^{14-10} = x^4, \frac{y^6}{y^4} = y^{6-4} =$

$y^2 \rightarrow \left(\frac{16x^7y^3}{48x^5y^2}\right)^2 = \frac{x^4y^2}{9}$

32) $\left(\frac{12x^5y^6}{60x^7y^2}\right)^2 = \frac{12^{1 \times 2} x^{5 \times 2} y^{6 \times 2}}{(12 \times 5)^{1 \times 2} x^{7 \times 2} y^{2 \times 2}} = \frac{12^2 x^{10} y^{12}}{12^2 \times 5^2 x^{14} y^4} = \frac{x^{10} y^{12}}{25x^{14} y^4} \rightarrow \frac{x^{10}}{x^{14}} = x^{10-14} = x^{-4} = \frac{1}{x^4}, \frac{y^{12}}{y^4} =$

$y^{12-4} = y^8 \rightarrow \left(\frac{12x^5y^6}{60x^7y^2}\right)^2 = \frac{y^8}{25x^4}$

33) $\left(\frac{1}{3}\right)^{-2} = 3^2 = 9$

34) $\left(\frac{1}{4}\right)^{-3} = 4^3 = 64$

35) $\left(\frac{1}{6}\right)^{-2} = 6^2 = 36$

36) $\left(\frac{2}{3}\right)^{-3} = \left(\frac{3}{2}\right)^3 = \frac{3^3}{2^3} = \frac{27}{8}$

37) $\left(\frac{2}{5}\right)^{-3} = \left(\frac{5}{2}\right)^3 = \frac{5^3}{2^3} = \frac{125}{8}$

38) $\left(\frac{3}{5}\right)^{-2} = \left(\frac{5}{3}\right)^2 = \frac{5^2}{3^2} = \frac{25}{9}$

39) $2y^{-3} = \frac{2}{y^3}$

40) $13y^{-5} = \frac{13}{y^5}$

41) $-20x^{-2} = -\frac{20}{x^2}$

42) $15a^{-2}b^3 = \frac{15b^3}{a^2}$

43) $23a^2b^{-4}c^{-8} = \frac{23a^2}{b^4c^8}$

44) $-4x^4y^{-2}2^{-7} = -\frac{4x^4}{y^2}$

45) $\frac{16y}{x^3y^{-4}} \rightarrow \frac{y}{y^{-4}} = y^{1-(-4)} = y^5 \rightarrow \frac{16y}{x^3y^{-4}} = \frac{16y^5}{x^3}$

46) $\frac{30a^{-3}b}{-100c^{-2}} \rightarrow -\frac{30\div10}{100\div10} = -\frac{3}{10} \rightarrow \frac{30a^{-3}b}{-100c^{-2}} = -\frac{3c^2b}{10a^3}$

47) $0.00518 = 5.18 \times 10^{-3}$

48) $0.000042 = 4.2 \times 10^{-5}$

49) $78,000 = 7.8 \times 10^4$

50) $92,000,000 = 9.2 \times 10^7$

51) $\sqrt{5} \times \sqrt{5} = 5$

52) $\sqrt{25} - \sqrt{4} = 5 - 2 = 3$

53) $\sqrt{81} + \sqrt{36} = 9 + 6 = 15$

54) $\sqrt{4} \times \sqrt{25} = 2 + 5 = 10$

55) $\sqrt{2} \times \sqrt{18} = \sqrt{36} = 6$

56) $4\sqrt{2} + 3\sqrt{2} = 7\sqrt{2}$

57) $5\sqrt{7} + 2\sqrt{7} = 7\sqrt{7}$

58) $\sqrt{45} + 2\sqrt{5} = \sqrt{5 \times 9} + 2\sqrt{5} = 3\sqrt{5} + 2\sqrt{5} = 5\sqrt{5}$

3 Expressions and Variables

Math topics that you'll learn in this chapter:

27

Simplifying Variable Expressions

☆ In algebra, a variable is a letter used to stand for a number. The most common letters are $x, y, z, a, b, c, m,$ and n.

☆ An algebraic expression is an expression that contains integers, variables, and math operations such as addition, subtraction, multiplication, division, etc.

☆ In an expression, we can combine "like" terms. (values with same variable and same power)

Examples:

Example 1. Simplify. $(3x + 9x + 2) =$

Solution: In this expression, there are three terms: $3x,\ 9x,$ and 2. Two terms are "like terms": $3x$ and $9x$. Combine like terms. $3x + 9x = 12x$. Then: $(3x + 9x + 2) = 12x + 2$ (***remember you cannot combine variables and numbers.***)

Example 2. Simplify. $-17x^2 + 6x + 15x^2 - 13 =$

Solution: Combine "like" terms: $-17x^2 + 15x^2 = -2x^2$
Then: $-17x^2 + 6x + 15x^2 - 13 = -2x^2 + 6x - 13$

Example 3. Simplify. $5x - 18 - 6x^2 + 3x^2 =$

Solution: Combine like terms. Then:
$5x - 18 - 6x^2 + 3x^2 = -3x^2 + 5x - 18$
Example 4. Simplify. $-5x - 4x^2 + 9x - 11x^2 =$
Solution: Combine "like" terms: $-5x + 9x = 4x$, and $-4x^2 - 11x^2 = -15x^2$
Then: $-5x - 4x^2 + 9x - 11x^2 = 4x - 15x^2$. Write in standard form (biggest powers first): $4x - 15x^2 = -15x^2 + 4x$

Simplifying Polynomial Expressions

☆ In mathematics, a polynomial is an expression consisting of variables and coefficients that involves only the operations of addition, subtraction, multiplication, and non–negative integer exponents of variables.

$$P(x) = a_n x^n + a_{n-1} x^{n-1} + \dots + a_2 x^2 + a_1 x + a_0$$

☆ Polynomials must always be simplified as much as possible. It means you must add together any like terms. (values with same variable and same power)

Examples:

Example 1. Simplify this Polynomial Expressions. $-2x^2 + 9x^3 + 5x^3 - 7x^4$

Solution: Combine "like" terms: $9x^3 + 5x^3 = 14x^3$

Then: $-2x^2 + 9x^3 + 5x^3 - 7x^4 = -2x^2 + 14x^3 - 7x^4$
Now, write the expression in standard form:
$-2x^2 + 14x^3 - 7x^4 = -7x^4 + 14x^3 - 2x^2$

Example 2. Simplify this expression. $(4x^2 - x^3) - (-6x^3 + 3x^2) =$

Solution: First, multiply $(-)$ into $(-6x^3 + 3x^2)$:

$(4x^2 - x^3) - (-6x^3 + 3x^2) = 4x^2 - x^3 + 6x^3 - 3x^2$
Then combine "like" terms: $4x^2 - x^3 + 6x^3 - 3x^2 = x^2 + 5x^3$
And write in standard form: $x^2 + 5x^3 = 5x^3 + x^2$

Example 3. Simplify. $-2x^3 + 6x^4 - 5x^2 - 14x^4 =$

Solution: Combine "like" terms: $6x^4 - 14x^4 = -8x^4$
Then: $-2x^3 + 6x^4 - 5x^2 - 14x^4 = -2x^3 - 8x^4 - 5x^2$
And write in standard form: $-2x^3 - 8x^4 - 5x^2 = -8x^4 - 2x^3 - 5x^2$

The Distributive Property

☆ The distributive property (or the distributive property of multiplication over addition and subtraction) simplifies and solves expressions in the form of: $a(b + c)$ or $a(b - c)$

☆ The distributive property is multiplying a term outside the parentheses by the terms inside.

☆ Distributive Property rule: $a(b + c) = ab + ac$

Examples:

Example 1. Simply using the distributive property. $(3)(4x - 9)$

Solution: Use Distributive Property rule: $a(b + c) = ab + ac$

$(3)(4x - 9) = (3 \times 4x) + (3) \times (-9) = 12x - 27$

Example 2. Simply. $(-4)(-3x + 8)$

Solution: Use Distributive Property rule: $a(b + c) = ab + ac$

$(-4)(-3x + 8) = (-4 \times (-3x)) + (-4) \times (8) = 12x - 32$

Example 3. Simply. $(5)(3x + 4) - 13x$

Solution: First, simplify $(5)(3x + 4)$ using the distributive property.

Then: $(5)(3x + 4) = 15x + 20$

Now combine like terms: $(5)(3x + 4) - 13x = 15x + 20 - 13x$

In this expression, $15x$ and $-13x$ are "like terms" and we can combine them.

$15x - 13x = 2x$. Then: $15x + 20 - 13x = 2x + 20$

www.EffortlessMath.com

Evaluating One Variable

☆ To evaluate one variable expressions, find the variable and substitute a number for that variable.

☆ Perform the arithmetic operations.

Examples:

Example 1. Calculate this expression for $x = 1$. $9 + 8x$

Solution: First, substitute 1 for x.

Then: $9 + 8x = 9 + 8(1)$

Now, use order of operation to find the answer: $9 + 8(1) = 9 + 8 = 17$

Example 2. Evaluate this expression for $x = -2$. $7x - 3$

Solution: First, substitute -2 for x.

Then: $7x - 3 = 7(-2) - 3$

Now, use order of operation to find the answer: $7(-2) - 3 = -14 - 3 = -17$

Example 3. Find the value of this expression when $x = 3$. $(12 - 2x)$

Solution: First, substitute 3 for x,

Then: $12 - 2x = 12 - 2(3) = 12 - 6 = 6$

Example 4. Solve this expression for $x = -4$. $11 + 5x$

Solution: Substitute -4 for x.

Then: $11 + 5x = 11 + 5(-4) = 11 - 20 = -9$

Evaluating Two Variables

☆ To evaluate an algebraic expression, substitute a number for each variable.

☆ Perform the arithmetic operations to find the value of the expression.

Examples:

Example 1. Calculate this expression for $a = -2$ and $b = 3$. $(2a - 6b)$

Solution: First, substitute -2 for a, and 3 for b.
Then: $2a - 6b = 2(-2) - 6(3)$
Now, use order of operation to find the answer: $2(-2) - 6(3) = -4 - 18 = -22$

Example 2. Evaluate this expression for $x = -3$ and $y = 4$. $(2x - 4y)$

Solution: Substitute -3 for x, and 4 for y.
Then: $2x - 4y = 2(-3) - 4(4) = -6 - 16 = -22$

Example 3. Find the value of this expression $3(-4a + 2b)$, when $a = -2$ and $b = -3$.

Solution: Substitute -2 for a, and -3 for b.
Then: $3(-4a + 2b) = 3\big(-4(-2) + 2(-3)\big) = 3(8 - 6) = 3(2) = 6$

Example 4. Evaluate this expression. $-5x - 3y$, $x = 2$, $y = -6$

Solution: Substitute 2 for x, and -6 for y and simplify.
Then: $-5x - 3y = -5(2) - 3(-6) = -10 + 18 = 8$

Day 3: Practices

✍ Simplify each expression.

1) $2 - 3x - 1 =$

2) $-6 - 2x + 8 =$

3) $11x - 6x - 4 =$

4) $-16x + 25x - 5 =$

5) $5x + 5 - 15x =$

6) $4 + 5x - 6x - 5 =$

7) $3x + 10 - 2x - 20 =$

8) $-3 - 2x^2 - 5 + 3x =$

9) $-7 + 9x^2 - 2 + 2x =$

10) $4x^2 + 2x - 12x - 5 =$

11) $2x^2 - 3x - 5x + 6 - 9 =$

12) $x^2 - 6x - x + 2 - 3 =$

13) $10x^2 - x - 8x + 3 - 10 =$

14) $4x^2 - 7x - x^2 + 2x + 5 =$

✍ Simplify each polynomial.

15) $4x^2 + 3x^3 - x^2 + x =$

16) $5x^4 + x^5 - x^4 + 4x^2 =$

17) $15x^3 + 12x - 6x^2 - 9x^3 =$

18) $(7x^3 - 2x^2) + (6x^2 - 13x) =$

19) $(9x^4 + 6x^3) + (11x^3 - 5x^4) =$

20) $(15x^5 - 5x^3) - (4x^3 + 6x^2) =$

21) $(15x^4 + 7x^3) - (3x^3 - 26) =$

22) $(22x^4 + 6x^3) - (-2x^3 - 4x^4) =$

23) $(x^2 + 6x^3) + (-19x^2 + 6x^3) =$

24) $(2x^4 - x^3) + (-5x^3 - 7x^4) =$

✎ **Use the distributive property to simply each expression.**

25) $3(5 + x) =$

26) $5(4 - x) =$

27) $6(2 - 5x) =$

28) $(4 - 3x)7 =$

29) $8(3 - 3x) =$

30) $(-1)(-6 + 2x) =$

31) $(-5)(3x - 3) =$

32) $(-x + 10)(-3)$

33) $(-2)(2 - 6x) =$

34) $(-6x - 4)(-7) =$

✎ **Evaluate each expression using the value given.**

35) $x = 3 \rightarrow 12 - x =$

36) $x = 5 \rightarrow x + 7 =$

37) $x = 3 \rightarrow 3x - 5 =$

38) $x = 2 \rightarrow 18 - 3x =$

39) $x = 7 \rightarrow 5x - 4 =$

40) $x = 6 \rightarrow 21 - x =$

41) $x = 5 \rightarrow 10x - 20 =$

42) $x = -5 \rightarrow 4 - x =$

43) $x = -2 \rightarrow 25 - 3x =$

44) $x = -7 \rightarrow 16 - x =$

45) $x = -13 \rightarrow 40 - 2x =$

46) $x = -4 \rightarrow 20x - 6 =$

47) $x = -6 \rightarrow -11x - 19 =$

48) $x = -8 \rightarrow -1 - 3x =$

✎ **Evaluate each expression using the values given.**

49) $x = 3, \; y = 2 \rightarrow 3x + 2y =$

50) $a = 4, \; b = 1 \rightarrow 2a - 6b =$

51) $x = 5, \; y = 7 \rightarrow 2x - 4y - 5 =$

52) $a = -3, \; b = 4 \rightarrow -3a + 4b + 2 =$

53) $x = -4, \; y = -3 \rightarrow 2x - 6 - 4y =$

Day 3: Answers

1) $2 - 3x - 1 \rightarrow 2 - 1 = 1 \rightarrow 2 - 3x - 1 = -3x + 1$

2) $-6 - 2x + 8 \rightarrow -6 + 8 = 2 \rightarrow -6 - 2x + 8 = -2x + 2$

3) $11x - 6x - 4 \rightarrow 11x - 6x = 5x \rightarrow 11x - 6x - 4 = 5x - 4$

4) $-16x + 25x - 5 \rightarrow -16x + 25x = 9x \rightarrow -16x + 25x - 5 = 9x - 5$

5) $5x + 5 - 15x \rightarrow 5x - 15x = -10x \rightarrow 5x + 5 - 15x = -10x + 5$

6) $4 + 5x - 6x - 5 \rightarrow 4 - 5 = -1, 5x - 6x = -x \rightarrow 4 + 5x - 6x - 5 = -x - 1$

7) $3x + 10 - 2x - 20 \rightarrow 10 - 20 = -10, 3x - 2x = x \rightarrow 3x + 10 - 2x - 20 = x - 10$

8) $-3 - 2x^2 - 5 + 3x \rightarrow -3 - 5 = -8 \rightarrow -3 - 2x^2 - 5 + 3x = -2x^2 + 3x - 8$

9) $-7 + 9x^2 - 2 + 2x \rightarrow -7 - 2 = -9 \rightarrow -7 + 9x^2 - 2 + 2x = 9x^2 + 2x - 9$

10) $4x^2 + 2x - 12x - 5 \rightarrow 2x - 12x = -10x \rightarrow 4x^2 + 2x - 12x - 5 = 4x^2 - 10x - 5$

11) $2x^2 - 3x - 5x + 6 - 9 \rightarrow -3x - 5x = -8x, 6 - 9 = -3 \rightarrow$

$2x^2 - 3x - 5x + 6 - 9 = 2x^2 - 8x - 3$

12) $x^2 - 6x - x + 2 - 3 \rightarrow -6x - x = -7x, 2 - 3 = -1 \rightarrow x^2 - 6x - x + 2 - 3 = $
$x^2 - 7x - 1$

13) $10x^2 - x - 8x + 3 - 10 \rightarrow -x - 8x = -9x, 3 - 10 = -7 \rightarrow 10x^2 - x - 8x + 3 -$
$10 = 10x^2 - 9x - 7$

14) $4x^2 - 7x - x^2 + 2x + 5 \rightarrow 4x^2 - x^2 = 3x^2, -7x + 2x = -5x \rightarrow$

$4x^2 - 7x - x^2 + 2x + 5 = 3x^2 - 5x + 5$

15) $4x^2 + 3x^3 - x^2 + x \rightarrow 4x^2 - x^2 = 3x^2 \rightarrow 4x^2 + 3x^3 - x^2 + x = 3x^3 + 3x^2 + x$

16) $5x^4 + x^5 - x^4 + 4x^2 \rightarrow 5x^4 - x^4 = 4x^4 \rightarrow 5x^4 + x^5 - x^4 + 4x^2 = $

$4x^4 + x^5 + 4x^2 = x^5 + 4x^4 + 4x^2$

17) $15x^3 + 12x - 6x^2 - 9x^3 \rightarrow 15x^3 - 9x^3 = 6x^3 \rightarrow 15x^3 + 12x - 6x^2 - 9x^3 =$

$6x^3 + 12x - 6x^2 = 6x^3 - 6x^2 + 12x$

18) $(7x^3 - 2x^2) + (6x^2 - 13x) = 7x^3 - 2x^2 + 6x^2 - 13x \rightarrow -2x^2 + 6x^2 = 4x^2 \rightarrow$

$7x^3 - 2x^2 + 6x^2 - 13x = 7x^3 + 4x^2 - 13x$

19) $(9x^4 + 6x^3) + (11x^3 - 5x^4) = 9x^4 + 6x^3 + 11x^3 - 5x^4 \rightarrow 9x^4 - 5x^4 = 4x^4,$

$6x^3 + 11x^3 = 17x^3 \rightarrow 9x^4 + 6x^3 + 11x^3 - 5x^4 = 4x^4 + 17x^3$

20) $(15x^5 - 5x^3) - (4x^3 + 6x^2) = 15x^5 - 5x^3 - 4x^3 - 6x^2 \rightarrow -5x^3 - 4x^3 = -9x^3 \rightarrow$

$15x^5 - 5x^3 - 4x^3 - 6x^2 = 15x^5 - 9x^3 - 6x^2$

21) $(15x^4 + 7x^3) - (3x^3 - 26) = 15x^4 + 7x^3 - 3x^3 + 26 \rightarrow 7x^3 - 3x^3 = 4x^3 \rightarrow$

$15x^4 + 7x^3 - 3x^3 + 26 = 15x^4 + 4x^3 + 26$

22) $(22x^4 + 6x^3) - (-2x^3 - 4x^4) = 22x^4 + 6x^3 + 2x^3 + 4x^4 \rightarrow 22x^4 + 4x^4 =$

$26x^4, 6x^3 + 2x^3 = 8x^3 \rightarrow 22x^4 + 6x^3 + 2x^3 + 4x^4 = 26x^4 + 8x^3$

23) $(x^2 + 6x^3) + (-19x^2 + 6x^3) = x^2 + 6x^3 - 19x^2 + 6x^3 \rightarrow 6x^3 + 6x^3 = 12x^3,$

$-19x^2 + x^2 = -18x^2 \rightarrow x^2 + 6x^3 - 19x^2 + 6x^3 = 12x^3 - 18x^2$

24) $(2x^4 - x^3) + (-5x^3 - 7x^4) = 2x^4 - x^3 - 5x^3 - 7x^4 \rightarrow 2x^4 - 7x^4 =$

$-5x^4, -x^3 - 5x^3 = -6x^3 \rightarrow 2x^4 - x^3 - 5x^3 - 7x^4 = -5x^4 - 6x^3$

25) $3(5 + x) = (3) \times (5) + (3) \times x = 15 + 3x = 3x + 15$

26) $5(4 - x) = (5) \times (4) + (5) \times (-x) = 20 + (-5x) = -5x + 20$

27) $6(2 - 5x) = (6) \times (2) + (6) \times (-5x) = 12 + (-30x) = -30x + 12$

28) $(4 - 3x)7 = (4) \times (7) + (-3x) \times (7) = 28 + (-21x) = -21x + 28$

29) $8(3 - 3x) = (8) \times (3) + (8) \times (-3x) = 24 + (-24x) = -24x + 24$

30) $(-1)(-6 + 2x) = (-1) \times (-6) + (-1) \times (2x) = 6 + (-2x) = -2x + 6$

31) $(-5)(3x - 3) = (-5) \times (3x) + (-5) \times (-3) = -15x + 15$

32) $(-x + 10)(-3) = (-x) \times (-3) + (10) \times (-3) = 3x - 30$

33) $(-2)(2-6x) = (-2) \times (2) + (-2) \times (-6x) = -4 + 12x = 12x - 4$

34) $(-6x-4)(-7) = (-6x) \times (-7) + (-4) \times (-7) = 42x + 28$

35) $x = 3 \rightarrow 12 - x = 12 - 3 = 9$

36) $x = 5 \rightarrow x + 7 = 5 + 7 = 12$

37) $x = 3 \rightarrow 3x - 5 = (3) \times (3) - 5 = 9 - 5 = 4$

38) $x = 2 \rightarrow 18 - 3x = 18 - (3) \times (2) = 18 - 6 = 12$

39) $x = 7 \rightarrow 5x - 4 = (5) \times (7) - 4 = 35 - 4 = 31$

40) $x = 6 \rightarrow 21 - x = 21 - 6 = 15$

41) $x = 5 \rightarrow 10x - 20 = (10) \times (5) - 20 = 50 - 20 = 30$

42) $x = -5 \rightarrow 4 - x = 4 - (-5) = 4 + 5 = 9$

43) $x = -2 \rightarrow 25 - 3x = 25 - (3) \times (-2) = 25 - (-6) = 25 + 6 = 31$

44) $x = -7 \rightarrow 16 - x = 16 - (-7) = 16 + 7 = 23$

45) $x = -13 \rightarrow 40 - 2x = 40 - (2) \times (-13) = 40 - (-26) = 40 + 26 = 66$

46) $x = -4 \rightarrow 20x - 6 = 20 \times (-4) - 6 = -80 - 6 = -86$

47) $x = -6 \rightarrow -11x - 19 = (-11) \times (-6) - 19 = 66 - 19 = 47$

48) $x = -8 \rightarrow -1 - 3x = (-1) - (3) \times (-8) = -1 - (-24) = -1 + 24 = 23$

49) $x = 3, \ y = 2 \rightarrow 3x + 2y = 3 \times (3) + 2(2) = 9 + 4 = 13$

50) $a = 4, \ b = 1 \rightarrow 2a - 6b = 2 \times (4) - 6(1) = 8 - 6 = 2$

51) $x = 5, \ y = 7 \rightarrow 2x - 4y - 5 = 2 \times (5) - 4(7) - 5 = 10 - 28 - 5 = -23$

52) $a = -3, \ b = 4 \rightarrow -3a + 4b + 2 = -3 \times (-3) + 4(4) + 2 = 9 + 16 + 2 = 27$

53) $x = -4, \ y = -3 \rightarrow 2x - 6 - 4y = 2 \times (-4) - 6 - 4(-3) = -8 - 6 + 12 = -2$

4 Equations and Inequalities

Math topics that you'll learn in this chapter:

39

One–Step Equations

☆ The values of two expressions on both sides of an equation are equal. Example: $ax = b$. In this equation, ax is equal to b.

☆ Solving an equation means finding the value of the variable.

☆ You only need to perform one Math operation to solve the one-step equations.

☆ To solve a one-step equation, find the inverse (opposite) operation is being performed.

☆ The inverse operations are:

- ❖ Addition and subtraction
- ❖ Multiplication and division

Examples:

Example 1. Solve this equation for x. $6x = 18 \rightarrow x = ?$

Solution: Here, the operation is multiplication (variable x is multiplied by 6) and its inverse operation is division. To solve this equation, divide both sides of equation by 6: $6x = 18 \rightarrow \frac{6x}{6} = \frac{18}{6} \rightarrow x = 3$

Example 2. Solve this equation. $x + 5 = 0 \rightarrow x = ?$

Solution: In this equation, 5 is added to the variable x. The inverse operation of addition is subtraction. To solve this equation, subtract 5 from both sides of the equation: $x + 5 - 5 = 0 - 5$. Then: $x + 5 - 5 = 0 - 5 \rightarrow x = -5$

Example 3. Solve this equation for x. $x - 11 = 0$

Solution: Here, the operation is subtraction and its inverse operation is addition. To solve this equation, add 11 to both sides of the equation: $x - 11 + 11 = 0 + 11 \rightarrow x = 11$

Multi–Step Equations

☆ To solve a multi-step equation, combine "like" terms on one side.

☆ Bring variables to one side by adding or subtracting.

☆ Simplify using the inverse of addition or subtraction.

☆ Simplify further by using the inverse of multiplication or division.

☆ Check your solution by plugging the value of the variable into the original equation.

Examples:

Example 1. Solve this equation for x. $5x - 6 = 26 - 3x$

Solution: First, bring variables to one side by adding $3x$ to both sides. Then:

$5x - 6 + 3x = 26 - 3x + 3x \rightarrow 5x - 6 + 3x = 26$.

Simplify: $8x - 6 = 26$. Now, add 6 to both sides of the equation:

$8x - 6 + 6 = 26 + 6 \rightarrow 8x = 32 \rightarrow$ Divide both sides by 8:

$8x = 32 \rightarrow \dfrac{8x}{8} = \dfrac{32}{8} \rightarrow x = 4$

Let's check this solution by substituting the value of 4 for x in the original equation:

$x = 4 \rightarrow 5x - 6 = 26 - 3x \rightarrow 5(4) - 6 = 26 - 3(4) \rightarrow 20 - 6 = 26 - 12 \rightarrow 14 = 14$

The answer $x = 4$ is correct.

Example 2. Solve this equation for x. $6x - 3 = 15$

Solution: Add 3 to both sides of the equation.

$6x - 3 = 15 \rightarrow 6x - 3 + 3 = 15 + 3 \rightarrow 6x = 18$

Divide both sides by 6, then: $6x = 18 \rightarrow \dfrac{6x}{6} = \dfrac{18}{6} \rightarrow x = 3$

Now, check the solution:

$x = 4 \rightarrow 6(3) - 3 = 15 \rightarrow 18 - 3 = 15 \rightarrow 15 = 15$

The answer $x = 4$ is correct.

System of Equations

★ A system of equations contains two equations and two variables. For example, consider the system of equations: $x - y = 1$ and $x + y = 5$

★ The easiest way to solve a system of equations is using the elimination method. The elimination method uses the addition property of equality. You can add the same value to each side of an equation.

★ For the first equation above, you can add $x + y$ to the left side and 5 to the right side of the first equation: $x - y + (x + y) = 1 + 5$. Now, if you simplify, you get: $x - y + (x + y) = 1 + 5 \rightarrow 2x = 6 \rightarrow x = 3$. Now, substitute 3 for the x in the first equation: $3 - y = 1$. By solving this equation, $y = 2$

Example:

What is the value of $x + y$ in this system of equations?

$$\begin{cases} -x + y = 18 \\ 2x - 6y = -12 \end{cases}$$

Solution: Solving the system of equations by elimination:
Multiply the first equation by (2), then add it to the second equation.

$\begin{array}{r} 2(-x + y = 18) \\ 2x - 6y = -12 \end{array} \Rightarrow \begin{array}{r} -2x + 2y = 36 \\ 2x - 6y = -12 \end{array} \Rightarrow (-2x) + 2x + 2y - 6y = 36 - 12 \Rightarrow -4y = 24 \Rightarrow$

$y = -6$

Plug in the value of y into one of the equations and solve for x.
$-x + (-6) = 18 \Rightarrow -x - 6 = 18 \Rightarrow -x = 24 \Rightarrow x = -24$
Thus, $x + y = -24 - 6 = -30$

Graphing Single–Variable Inequalities

☆ An inequality compares two expressions using an inequality sign.

☆ Inequality signs are: "less than" <, "greater than" >, "less than or equal to" ≤, and "greater than or equal to" ≥.

☆ To graph a single–variable inequality, find the value of the inequality on the number line.

☆ For less than (<) or greater than (>) draw an open circle on the value of the variable. If there is an equal sign too, then use a filled circle.

☆ Draw an arrow to the right for greater or to the left for less than.

Examples:

Draw a graph for this inequality. $x > 3$

Solution: Since the variable is greater than 3, then we need to find 3 in the number line and draw an open circle on it. Then, draw an arrow to the right.

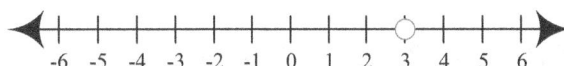

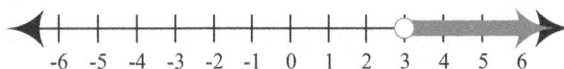

Graph this inequality. $x \leq -1$.

Solution: Since the variable is less than or equal to −1, then we need to find −1 on the number line and draw a filled circle on it. Then, draw an arrow to the left.

One–Step Inequalities

☆ An inequality compares two expressions using an inequality sign.

☆ Inequality signs are: "less than" $<$, "greater than" $>$, "less than or equal to" $\leq$, and "greater than or equal to" $\geq$.

☆ You only need to perform one Math operation to solve the one-step inequalities.

☆ To solve one-step inequalities, find the inverse (opposite) operation is being performed.

☆ For dividing or multiplying both sides by negative numbers, flip the direction of the inequality sign.

Examples:

Example 1. Solve this inequality for x. $x + 7 \geq 2$

Solution: The inverse (opposite) operation of addition is subtraction. In this inequality, 7 is added to x. To isolate x we need to subtract 7 from both sides of the inequality.

Then: $x + 7 \geq 2 \rightarrow x + 7 - 7 \geq 2 - 7 \rightarrow x \geq -5$. The solution is: $x \geq -5$

Example 2. Solve the inequality. $x - 2 > -12$

Solution: 2 is subtracted from x. Add 2 to both sides.

$x - 2 > -12 \rightarrow x - 2 + 2 > -12 + 2 \rightarrow x > -10$

Example 3. Solve. $6x \leq -36$

Solution: 6 is multiplied to x. Divide both sides by 6.

Then: $6x \leq -36 \rightarrow \frac{6x}{6} \leq \frac{-36}{6} \rightarrow x \leq -6$

Example 4. Solve. $-2x \leq 10$

Solution: -2 is multiplied to x. Divide both sides by -2. Remember when dividing or multiplying both sides of an inequality by negative numbers, flip the direction of the inequality sign.

Then: $-2x \leq 10 \rightarrow \frac{-2x}{-2} \geq \frac{10}{-2} \rightarrow x \geq -5$

Multi–Step Inequalities

☆ To solve a multi-step inequality, combine "like" terms on one side.

☆ Bring variables to one side by adding or subtracting.

☆ Isolate the variable.

☆ Simplify using the inverse of addition or subtraction.

☆ Simplify further by using the inverse of multiplication or division.

☆ For dividing or multiplying both sides by negative numbers, flip the direction of the inequality sign.

Examples:

Example 1. Solve this inequality. $4x - 1 \leq 23$

Solution: In this inequality, 1 is subtracted from $4x$. The inverse of subtraction is addition. Add 1 to both sides of the inequality:

$4x - 1 + 1 \leq 23 + 1 \rightarrow 4x \leq 24$

Now, divide both sides by 4. Then: $4x \leq 24 \rightarrow \frac{4x}{4} \leq \frac{24}{4} \rightarrow x \leq 6$

The solution of this inequality is $x \leq 6$.

Example 2. Solve this inequality. $2x - 6 < 18$

Solution: First, add 6 to both sides: $2x - 6 + 6 < 18 + 6$

Then simplify: $2x - 6 + 6 < 18 + 6 \rightarrow 2x < 24$

Now divide both sides by 2: $\frac{2x}{2} < \frac{24}{2} \rightarrow x < 12$

Example 3. Solve this inequality. $-4x - 8 \geq 12$

Solution: First, add 8 to both sides:

$-4x - 8 + 8 \geq 12 + 8 \rightarrow -4x \geq 20$

Divide both sides by -4. Remember that you need to flip the direction of inequality sign. $-4x \geq 20 \rightarrow \frac{-4x}{-4} \leq \frac{20}{-4} \rightarrow x \leq -5$

Day 4: Practices

✍ Solve each equation. (One–Step Equations)

1) $x + 2 = 5 \rightarrow x =$

2) $8 = 13 + x \rightarrow x =$

3) $-6 = 7 + x \rightarrow x =$

4) $x - 5 = -3 \rightarrow x =$

5) $-13 = x - 15 \rightarrow x =$

6) $-10 + x = -4 \rightarrow x =$

7) $-19 + x = 7 \rightarrow x =$

8) $-6x = 24 \rightarrow x =$

9) $\frac{x}{4} = -5 \rightarrow x =$

10) $-2x = -4 \rightarrow x =$

✍ Solve each equation. (Multi–Step Equations)

11) $2(x + 5) = 16 \rightarrow x =$

12) $-6(3 - x) = 18 \rightarrow x =$

13) $25 = -5(x + 4) \rightarrow x =$

14) $-12 = 6(9 + x) \rightarrow x =$

15) $11(x + 5) = -22 \rightarrow x =$

16) $-27 - 36x = 45 \rightarrow x =$

17) $3x - 4 = x - 12 \rightarrow x =$

18) $-8x + x - 11 = 24 \rightarrow x =$

✍ Solve each system of equations.

19) $\begin{cases} x + 4y = 29 \\ x + 2y = 5 \end{cases}$ $x =$ ____ $y =$ ____

20) $\begin{cases} 2x + y = 36 \\ x + 4y = 4 \end{cases}$ $x =$ ____ $y =$ ____

21) $\begin{cases} 2x + 5y = 15 \\ x + y = 6 \end{cases}$ $x =$ ____ $y =$ ____

22) $\begin{cases} 2x - 2y = -16 \\ -9x + 2y = -19 \end{cases}$ $x =$ ____ $y =$ ____

✍ Draw a graph for each inequality.

23) $x \leq 1$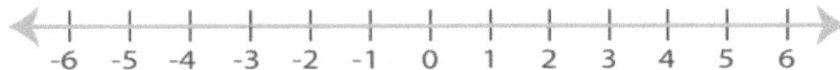

24) $x > -4$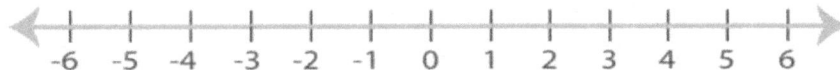

✍ Solve each inequality and graph it.

25) $x - 3 \geq -2$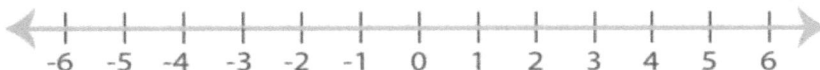

26) $7x - 6 < 8$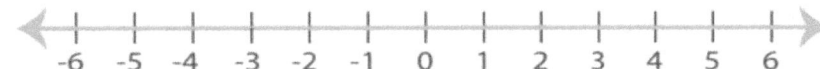

✍ Solve each inequality.

27) $x + 11 > 3$

28) $x + 4 > 1$

29) $-6 + 3x \leq 21$

30) $-5 + 4x \leq 19$

31) $4 + 9x \leq 31$

32) $8(x + 3) \geq -16$

33) $3(6 + x) \geq 18$

34) $3(x - 2) < -9$

35) $15 + 9x < -30$

36) $3(6 - x) \geq -27$

37) $4(x - 5) \geq -32$

38) $6(x + 4) < -24$

39) $7(x - 8) \geq -49$

40) $-(-6 - 5x) > -39$

41) $2(1 - 2x) > -66$

42) $-3(3 - 2x) > -33$

Day 4: Answers

1) $x + 2 = 5 \rightarrow x = 5 - 2 = 3$

2) $8 = 13 + x \rightarrow x = 8 - 13 = -5$

3) $-6 = 7 + x \rightarrow x = -6 - 7 = -13$

4) $x - 5 = -3 \rightarrow x = -3 + 5 = 2$

5) $-13 = x - 15 \rightarrow x = -13 + 15 = 2$

6) $-10 + x = -4 \rightarrow x = -4 + 10 = 6$

7) $-19 + x = 7 \rightarrow x = 7 + 19 = 26$

8) $-6x = 24 \rightarrow x = \frac{24}{-6} = -4$

9) $\frac{x}{4} = -5 \rightarrow x = -5 \times 4 = -20$

10) $-2x = -4 \rightarrow x = \frac{-4}{-2} = 2$

11) $2(x + 5) = 16 \rightarrow \frac{2(x+5)}{2} = \frac{16}{2} \rightarrow (x + 5) = 8 \rightarrow x = 8 - 5 = 3$

12) $-6(3 - x) = 18 \rightarrow \frac{-6(3-x)}{-6} = \frac{18}{-6} \rightarrow (3 - x) = -3 \rightarrow x = 3 + 3 = 6$

13) $25 = -5(x + 4) \rightarrow \frac{25}{-5} = \frac{-5(x+4)}{-5} \rightarrow -5 = (x + 4) \rightarrow x = -5 - 4 = -9$

14) $-12 = 6(9 + x) \rightarrow \frac{-12}{6} = \frac{6(9+x)}{6} \rightarrow -2 = (9 + x) \rightarrow x = -2 - 9 = -11$

15) $11(x + 5) = -22 \rightarrow \frac{11(x+5)}{11} = \frac{-22}{11} \rightarrow (x + 5) = -2 \rightarrow x = -2 - 5 = -7$

16) $-27 - 36x = 45 \rightarrow -36x = 45 + 27 \rightarrow -36x = 72 \rightarrow \frac{-36x}{-2} = \frac{72}{-2} \rightarrow x = -2$

17) $3x - 4 = x - 12 \rightarrow 3x - 4 - x = x - 12 - x \rightarrow 2x - 4 = -12 \rightarrow 2x = -12 + 4 \rightarrow$

   $2x = -8 \rightarrow \frac{2x}{2} = \frac{-8}{2} \rightarrow x = -4$

18) $-8x + x - 11 = 24 \rightarrow -7x = 24 + 11 \rightarrow \frac{-7x}{-7} = \frac{35}{-7} \rightarrow x = -5$

19) $\begin{cases} x + 4y = 29 \\ x + 2y = 5 \end{cases} \rightarrow \begin{matrix} -(x + 4y = 29) \\ x + 2y = 5 \end{matrix} \rightarrow \begin{matrix} -x - 4y = -29 \\ x + 2y = 5 \end{matrix} \rightarrow (-x) + x - 4y + 2y =$

   $-29 + 5 \rightarrow -2y = -24 \rightarrow \frac{-2y}{-2} = \frac{-24}{-2} \rightarrow y = 12$

   Plug in the value of y into one of the equations and solve for x.

   $x + 2y = 5 \rightarrow x + 2(12) = 5 \rightarrow x + 24 = 5 \rightarrow x = 5 - 24 = -19$

20) $\begin{cases} 2x + y = 36 \\ x + 4y = 4 \end{cases} \rightarrow \begin{matrix} 2x + y = 36 \\ -2(x + 4y = 4) \end{matrix} \rightarrow \begin{matrix} 2x + y = 36 \\ -2x - 8y = -8 \end{matrix} \rightarrow 2x - 2x + y - 8y = 36 -$

$8 \rightarrow -7y = 28 \rightarrow \frac{-7y}{-7} = \frac{28}{-7} \rightarrow y = -4$

Plug in the value of y into one of the equations and solve for x.

$x + 4y = 4 \rightarrow x + 4(-4) = 4 \rightarrow x - 16 = 4 \rightarrow x = 4 + 16 = 20$

21) $\begin{cases} 2x + 5y = 15 \\ x + y = 6 \end{cases} \rightarrow \begin{matrix} 2x + 5y = 15 \\ -2(x + y = 6) \end{matrix} \rightarrow \begin{matrix} 2x + 5y = 15 \\ -2x - 2y = -12 \end{matrix} \rightarrow 2x - 2x + 5y - 2y = 15 -$

$12 \rightarrow 3y = 3 \rightarrow \frac{3y}{3} = \frac{3}{3} \rightarrow y = 1$

Plug in the value of y into one of the equations and solve for x.

$x + y = 6 \rightarrow x + 1 = 6 \rightarrow x = 6 - 1 = 5$

22) $\begin{cases} 2x - 2y = -16 \\ -9x + 2y = -19 \end{cases} \rightarrow 2x - 9x - 2y + 2y = -16 - 19 \rightarrow -7x = -35 \rightarrow \frac{-7x}{-7} =$

$\frac{-35}{-7} \rightarrow x = 5$

Plug in the value of x into one of the equations and solve for y.

$2x - 2y = -16 \rightarrow 2(5) - 2y = -16 \rightarrow 10 - 2y = -16 \rightarrow 10 + 16 = 2y \rightarrow 26 = 2y$

$\rightarrow \frac{26}{2} = \frac{2y}{2} \rightarrow y = 13$

23) $x \leq 1$

24) $x > -4$

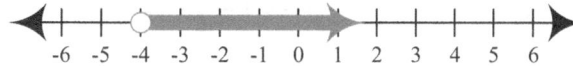

25) $x - 3 \geq -2 \rightarrow x \geq -2 + 3 \rightarrow x \geq 1$

26) $7x - 6 < 8 \rightarrow 7x < 8 + 6 \rightarrow 7x < 14 \rightarrow x < \frac{14 \div 7}{7 \div 7} \rightarrow x < 2$

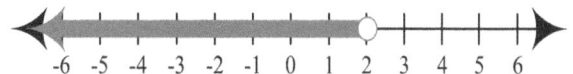

27) $x + 11 > 3 \rightarrow x > 3 - 11 \rightarrow x > -8$

28) $x + 4 > 1 \rightarrow x > 1 - 4 \rightarrow x > -3$

29) $-6 + 3x \leq 21 \rightarrow 3x \leq 21 + 6 \rightarrow 3x \leq 27 \rightarrow x \leq \frac{27 \div 3}{3 \div 3} \rightarrow x \leq 9$

30) $-5 + 4x \leq 19 \rightarrow 4x \leq 19 + 5 \rightarrow 4x \leq 24 \rightarrow x \leq \frac{24 \div 4}{4 \div 4} \rightarrow x \leq 6$

31) $4 + 9x \leq 31 \rightarrow 9x \leq 31 - 4 \rightarrow 9x \leq 27 \rightarrow x \leq \frac{27 \div 9}{9 \div 9} \rightarrow x \leq 3$

32) $8(x + 3) \geq -16 \rightarrow x + 3 \geq \frac{-16}{8} \rightarrow x + 3 \geq -2 \rightarrow x \geq -2 - 3 \rightarrow x \geq -5$

33) $3(6 + x) \geq 18 \rightarrow 6 + x \geq \frac{18}{3} \rightarrow 6 + x \geq 6 \rightarrow x \geq 6 - 6 \rightarrow x \geq 0$

34) $3(x - 2) < -9 \rightarrow x - 2 < \frac{-9}{3} \rightarrow x - 2 < -3 \rightarrow x < -3 + 2 \rightarrow x < -1$

35) $15 + 9x < -30 \rightarrow 9x < -30 - 15 \rightarrow 9x < -45 \rightarrow x < \frac{-45}{9} \rightarrow x < -5$

36) $3(6 - x) \geq -27 \rightarrow 6 - x \geq \frac{-27}{3} \rightarrow 6 - x \geq -9 \rightarrow -x \geq -9 - 6 \rightarrow \frac{-x}{-1} \leq \frac{-15}{-1} \rightarrow$

$\quad x \leq 15$

37) $4(x - 5) \geq -32 \rightarrow x - 5 \geq \frac{-32}{4} \rightarrow x - 5 \geq -8 \rightarrow x \geq -8 + 5 \rightarrow x \geq -3$

38) $6(x + 4) < -24 \rightarrow x + 4 < \frac{-24}{6} \rightarrow x + 4 < -4 \rightarrow x < -4 - 4 \rightarrow x < -8$

39) $7(x - 8) \geq -49 \rightarrow x - 8 \geq \frac{-49}{7} \rightarrow x - 8 \geq -7 \rightarrow x \geq -7 + 8 \rightarrow x \geq 1$

40) $-(-6 - 5x) > -39 \rightarrow 6 + 5x > -39 \rightarrow 5x > -39 - 6 \rightarrow 5x > -45 \rightarrow$

$\quad x > \frac{-45}{5} \rightarrow x > -9$

41) $2(1 - 2x) > -66 \rightarrow 1 - 2x > \frac{-66}{2} \rightarrow 1 - 2x > -33 \rightarrow -2x > -33 - 1 \rightarrow$

$\quad -2x > -34 \rightarrow x < \frac{-34}{-2} \rightarrow x < \frac{-34}{-2} \rightarrow x < 17$

42) $-3(3 - 2x) > -33 \rightarrow -9 + 6x > -33 \rightarrow 6x > -33 + 9 \rightarrow 6x > -24 \rightarrow$

$\quad \frac{6x}{6} > \frac{-24}{6} \rightarrow x > -4$

5 Lines and Slope

Math topics that you'll learn in this chapter:

51

Finding Slope

☆ The slope of a line represents the direction of a line on the coordinate plane.

☆ A coordinate plane contains two perpendicular number lines. The horizontal line is x and the vertical line is y. The point at which the two axes intersect is called the origin. An ordered pair (x, y) shows the location of a point.

☆ A line on a coordinate plane can be drawn by connecting two points.

☆ To find the slope of a line, we need the equation of the line or two points on the line.

☆ The slope of a line with two points A (x_1, y_1) and B (x_2, y_2) can be found by using this formula: $\frac{y_2 - y_1}{x_2 - x_1} = \frac{rise}{run}$

☆ The equation of a line is typically written as $y = mx + b$ where m is the slope and b is the y-intercept.

Examples:

Example 1. Find the slope of the line through these two points:

$A(5, -5)$ and $B(9, 7)$.

Solution: Slope $= \frac{y_2 - y_1}{x_2 - x_1}$. Let (x_1, y_1) be A$(5, -5)$ and (x_2, y_2) be $B(9, 7)$.

(Remember, you can choose any point for (x_1, y_1) and (x_2, y_2)).

Then: slope $= \frac{y_2 - y_1}{x_2 - x_1} = \frac{7 - (-5)}{9 - 5} = \frac{12}{4} = 3$

The slope of the line through these two points is 3.

Example 2. Find the slope of the line with equation $y = -2x + 8$

Solution: When the equation of a line is written in the form of $y = mx + b$, the slope is m. In this line: $y = -2x + 8$, the slope is -2.

Graphing Lines Using Slope–Intercept Form

☆ Slope–intercept form of a line: given the slope m and the y-intercept (the intersection of the line and y-axis) b, then the equation of the line is:

$$y = mx + b$$

☆ To draw the graph of a linear equation in a slope-intercept form on the xy coordinate plane, find two points on the line by plugging two values for x and calculating the values of y.

☆ You can also use the slope (m) and one point to graph the line.

Example:

Sketch the graph of $y = -3x + 6$.

Solution: To graph this line, we need to find two points. When x is zero the value of y is 6. And when y is 0 the value of x is 2.

$$x = 0 \rightarrow y = -3(0) + 6 = 6$$
$$y = 0 \rightarrow 0 = -3x + 6 \rightarrow x = 2$$

Now, we have two points:
$(0, 6)$ and $(2, 0)$.
Find the points on the coordinate plane and graph the line. Remember that the slope of the line is -3.

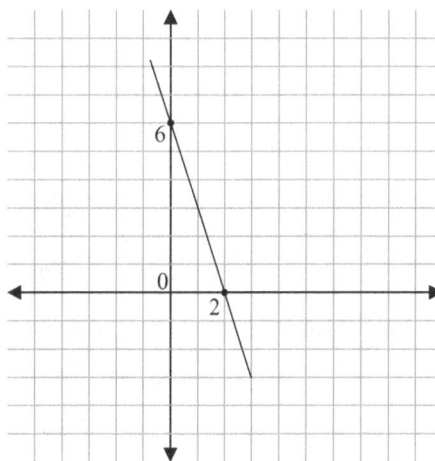

Writing Linear Equations

☆ The equation of a line in slope-intercept form: $y = mx + b$

☆ To write the equation of a line, first identify the slope.

☆ Find the y-intercept. This can be done by substituting the slope and the coordinates of a point (x, y) on the line.

Examples:

Example 1. What is the equation of the line that passes through $(-7, 2)$ and has a slope of 4?

Solution: The general slope-intercept form of the equation of a line is $y = mx + b$, where m is the slope and b is the y-intercept.
By substitution of the given point and given slope:
$y = mx + b \rightarrow 2 = (4)(-7) + b$. So, $b = 2 + 28 = 30$, and the required equation of the line is: $y = 4x + 30$

Example 2. Write the equation of the line through two points $A(5, 2)$ and $B(3, -4)$.

Solution: First, find the slope:
$$Slop = \frac{y_2 - y_1}{x_2 - x_1} = \frac{-4 - 2}{3 - 5} = \frac{-6}{-2} = 3 \rightarrow m = 3$$
To find the value of b, use either points and plug in the values of x and y in the equation. The answer will be the same: $y = x + b$. Let's check both points. Then:
$(5, 2) \rightarrow y = mx + b \rightarrow 2 = 3(5) + b \rightarrow b = -13$
$(3, -4) \rightarrow y = mx + b \rightarrow -4 = 3(3) + b \rightarrow b = -13$.
The y-intercept of the line is -13. The equation of the line is: $y = 3x - 13$

Finding Midpoint

☆ The middle of a line segment is its midpoint.

☆ The Midpoint of two endpoints A (x_1, y_1) and B (x_2, y_2) can be found using this formula: $M = \left(\frac{x_1+x_2}{2}, \frac{y_1+y_2}{2}\right)$

Examples:

Example 1. Find the midpoint of the line segment with the given endpoints. $(3, 5), (1, 3)$

Solution: Midpoint $= \left(\frac{x_1+x_2}{2}, \frac{y_1+y_2}{2}\right) \rightarrow (x_1, y_1) = (3, 5)$ and $(x_2, y_2) = (1, 3)$
Midpoint $= \left(\frac{3+1}{2}, \frac{5+3}{2}\right) \rightarrow \left(\frac{4}{2}, \frac{8}{2}\right) \rightarrow M(2, 4)$

Example 2. Find the midpoint of the line segment with the given endpoints. $(-1, 3), (9, -9)$

Solution: Midpoint $= \left(\frac{x_1+x_2}{2}, \frac{y_1+y_2}{2}\right) \rightarrow (x_1, y_1) = (-1, 3)$ and $(x_2, y_2) = (9, -9)$
Midpoint $= \left(\frac{-1+9}{2}, \frac{3+(-9)}{2}\right) \rightarrow \left(\frac{8}{2}, \frac{-6}{2}\right) \rightarrow M(4, -3)$

Example 3. Find the midpoint of the line segment with the given endpoints. $(8, 4), (-2, 6)$

Solution: Midpoint $= \left(\frac{x_1+x_2}{2}, \frac{y_1+y_2}{2}\right) \rightarrow (x_1, y_1) = (8, 4)$ and $(x_2, y_2) = (-2, 6)$
Midpoint $= \left(\frac{8-2}{2}, \frac{4+6}{2}\right) \rightarrow \left(\frac{6}{2}, \frac{10}{2}\right) \rightarrow M(3, 5)$

Example 4. Find the midpoint of the line segment with the given endpoints. $(7, -4), (-3, -8)$

Solution: Midpoint $= \left(\frac{x_1+x_2}{2}, \frac{y_1+y_2}{2}\right) \rightarrow (x_1, y_1) = (7, -4)$ and $(x_2, y_2) = (-3, -8)$
Midpoint $= \left(\frac{7-3}{2}, \frac{-4-8}{2}\right) \rightarrow \left(\frac{4}{2}, \frac{-12}{2}\right) \rightarrow M(2, -6)$

Finding Distance of Two Points

☆ Use the following formula to find the distance of two points with the coordinates A (x_1, y_1) and B (x_2, y_2):

$$d = \sqrt{(x_2 - x_1)^2 + (y_2 - y_1)^2}$$

Examples:

Example 1. Find the distance between $(5, -6)$ and $(-3, 9)$. on the coordinate plane.

Solution: Use distance of two points formula: $d = \sqrt{(x_2 - x_1)^2 + (y_2 - y_1)^2}$

$(x_1, y_1) = (5, -6)$ and $(x_2, y_2) = (-3, 9)$. Then: $d = \sqrt{(x_2 - x_1)^2 + (y_2 - y_1)^2} =$

$\sqrt{(-3 - 5)^2 + (9 - (-6))^2} = \sqrt{(-8)^2 + (15)^2} = \sqrt{64 + 225} = \sqrt{289} = 17$.

Then: $d = 17$

Example 2. Find the distance of two points $(-3, 10)$ and $(-9, 2)$

Solution: Use distance of two points formula: $d = \sqrt{(x_2 - x_1)^2 + (y_2 - y_1)^2}$

$(x_1, y_1) = (\text{-}3, 10)$ and $(x_2, y_2) = (-9, 2)$

Then: $d = \sqrt{(x_2 - x_1)^2 + (y_2 - y_1)^2} \rightarrow d = \sqrt{(-9 - (-3))^2 + (2 - 10)^2} =$

$\sqrt{(-6)^2 + (-8)^2} = \sqrt{36 + 64} = \sqrt{100} = 10$. Then: $d = 10$

Example 3. Find the distance between $(-8, 7)$ and $(4, -9)$.

Solution: Use distance of two points formula: $d = \sqrt{(x_2 - x_1)^2 + (y_2 - y_1)^2}$

$(x_1, y_1) = (-8, 7)$ and $(x_2, y_2) = (4, -9)$. Then: $d = \sqrt{(x_2 - x_1)^2 + (y_2 - y_1)^2}$

$d = \sqrt{\left(4 - (-8)\right)^2 + (-9 - 7)^2} = \sqrt{(12)^2 + (-16)^2} = \sqrt{144 + 256} = \sqrt{400} = 20$.

Then: $d = 20$

Graphing Linear Inequalities

☆ To graph a linear inequality, first draw a graph of the "equals" line.

☆ Use a dash line for less than (<) and greater than (>) signs and a solid line for less than and equal to (≤) and greater than and equal to (≥).

☆ Choose a testing point. (it can be any point on both sides of the line.)

☆ Put the value of (x, y) of that point in the inequality. If that works, that part of the line is the solution. If the values don't work, then the other part of the line is the solution.

Example:

Sketch the graph of inequality: $y > 2x - 5$

Solution: To draw the graph of $y > 2x - 5$, you first need to graph the line:

$y = 2x - 5$

Since there is a greater than (>) sign, draw a dash line.

The slope is 2 and y-intercept is -5.

Then, choose a testing point and substitute the value of x and y from that point into the inequality. The easiest point to test is the origin: $(0, 0)$

$(0, 0) \rightarrow y > 2x - 5 \rightarrow 0 > 2(0) - 5 \rightarrow 0 > -5$

This is correct! 0 is greater than -5. So, this part of the line (on the left side) is the solution of this inequality.

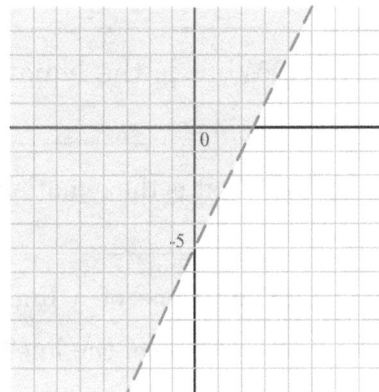

Day 7: Practices

✍ **Find the slope of each line.**

1) $y = x - 3$

2) $y = 3x + 4$

3) $y = -2x + 4$

4) Line through $(2, 5)$ and $(3, -4)$

5) Line through $(0, 6)$ and $(2, 4)$

6) Line through $(-2, 4)$ and $(3, -6)$

✍ **Sketch the graph of each line. (Using Slope–Intercept Form)**

7) $y = x + 3$

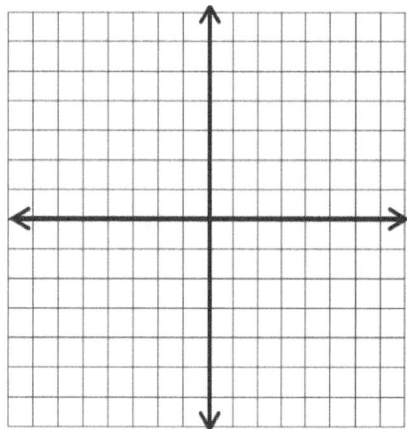

8) $y = x - 3$

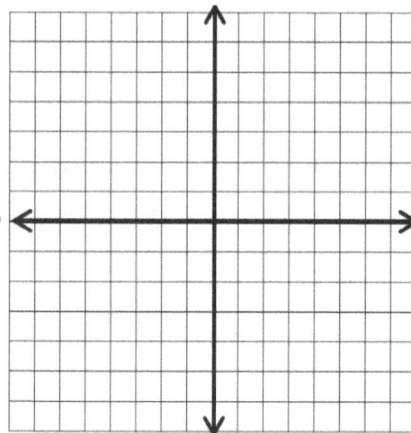

✍ **Solve.**

9) What is the equation of a line with slope 3 and intercept 12?

_____

10) What is the equation of a line with slope 4 and passes through point $(2, 4)$? _____

11) What is the equation of a line with slope −2 and passes through point $(5, -3)$?

_____

12) The slope of a line is −5 and it passes through point $(-4, 3)$. What is the equation of the line? _____

13) The slope of a line is −6 and it passes through point $(-2, -3)$. What is the equation of the line? _____

✍ **Sketch the graph of each linear inequality.**

14) $y > 3x - 3$

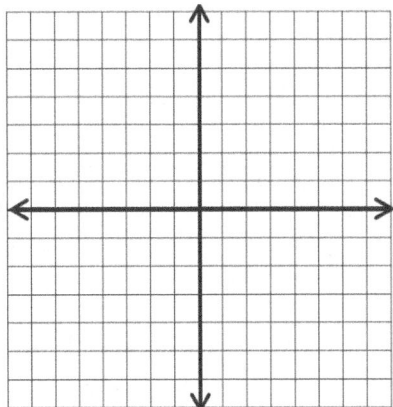

15) $y > -2x + 1$

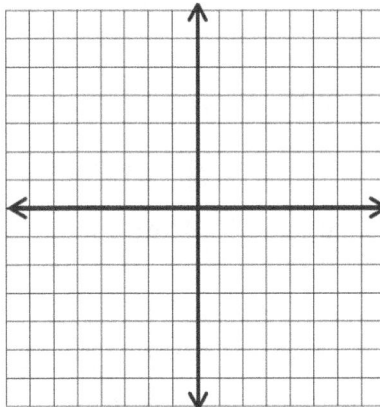

✍ **Find the midpoint of the line segment with the given endpoints.**

16) $(4, 1), (2, 3)$

17) $(3, 6), (5, 4)$

18) $(7, 1), (1, 3)$

19) $(2, 8), (2, 10)$

20) $(3, -2), (-1, 6)$

21) $(-1, -3), (1, 5)$

22) $(1, 4), (-7, 6)$

23) $(-3, 5), (7, -9)$

✍ **Find the distance between each pair of points.**

24) $(-8, -1), (-4, 2)$

25) $(-15, 2), (5, -13)$

26) $(-1, 11), (-7, 3)$

27) $(0, 11), (9, 11)$

28) $(-2, 4), (3, -8)$

29) $(6, -7), (-9, 1)$

30) $(8, -4), (-4, -20)$

31) $(5, 1), (9, -2)$

32) $(-8, -17), (2, 7)$

33) $(18, 21), (-12, 5)$

Day 5: Answers

1) $y = mx + b$, the slope is m. In this line: $y = x - 3$, the slope is $m = 1$.

2) $y = 3x + 4$, the slope is $m = 3$.

3) $y = -2x + 4$, the slope is $m = -2$.

4) $(x_1, y_1) = (2, 5)$ and $(x_2, y_2) = (3, 4) \rightarrow m = \frac{y_2 - y_1}{x_2 - x_1} = \frac{4-5}{3-2} = \frac{-1}{1} = -1$

5) $(x_1, y_1) = (0, 6)$ and $(x_2, y_2) = (2, -4) \rightarrow m = \frac{y_2 - y_1}{x_2 - x_1} = \frac{-4-6}{2-0} = \frac{-10}{2} = -5$

6) $(x_1, y_1) = (-2, 4)$ and $(x_2, y_2) = (3, -6) \rightarrow m = \frac{y_2 - y_1}{x_2 - x_1} = \frac{-6-4}{3-(-2)} = \frac{-6-4}{3+2} = \frac{-10}{5} = -2$

7) $y = x + 3$

$x = 0 \rightarrow y = 0 + 3 = 3 \rightarrow (0, 3)$

$y = 0 \rightarrow 0 = x + 3 \rightarrow x = -3$
$\rightarrow (-3, 0)$

$x = 1 \rightarrow y = 1 + 3 = 4 \rightarrow (1, 4)$

$y = 1 \rightarrow 1 = x + 3 \rightarrow x = 1 - 3$
$= -2 \rightarrow (-2, 1)$

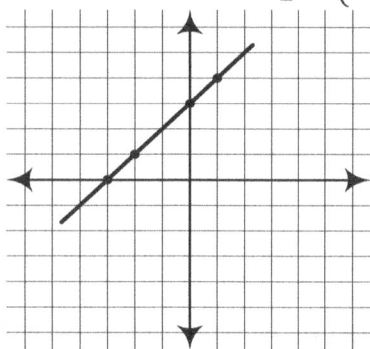

8) $y = x - 3$

$x = 0 \rightarrow y = 0 - 3 = -3 \rightarrow (0, -3)$

$y = 0 \rightarrow 0 = x - 3 \rightarrow x = 3 \rightarrow$
$\rightarrow (3, 0)$

$x = 1 \rightarrow y = 1 - 3 = -2 \rightarrow (1, -2)$

$y = 1 \rightarrow 1 = x - 3 \rightarrow x = 1 + 3 = 4$
$\rightarrow (4, 1)$

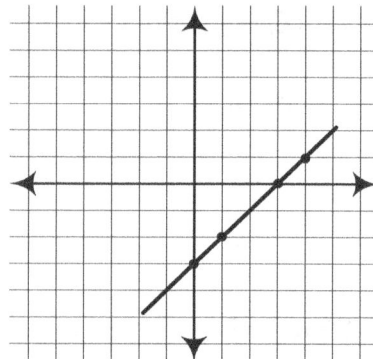

9) The general slope-intercept form of the equation of a line is $y = mx + b$, where m is the slope and b is the y-intercept $\rightarrow y = 3x + 12$

10) $y = mx + b \rightarrow 4 = 4(2) + b \rightarrow 4 = 8 + b \rightarrow b = 4 - 8 = -4 \rightarrow y = 4x - 4$

11) $y = mx + b \rightarrow -3 = -2(5) + b \rightarrow -3 = -10 + b \rightarrow b = -3 + 10 = 7 \rightarrow y = -2x + 7$

12) $y = mx + b \rightarrow 3 = -5(-4) + b \rightarrow 3 = 20 + b \rightarrow b = 3 - 20 = -17 \rightarrow y = -5x - 17$

13) $y = mx + b \rightarrow -3 = -6(-2) + b \rightarrow -3 = 12 + b \rightarrow b = -3 - 12 = -15 \rightarrow y = -6x - 15$

14) $y > 3x - 3$

$x = 0 \rightarrow y = 0 - 3 = -3 \rightarrow (0, -3)$

$y = 0 \rightarrow 0 = 3x - 3 \rightarrow 3x = 3 \rightarrow x = 1 \rightarrow (1, 0)$

The easiest point to test is the origin: $(0, 0)$

$(0,0) \rightarrow y > 3x - 3 \rightarrow 0 > 3(0) - 3 \rightarrow 0 > -3$

This is correct! 0 is greater than -3. So, this part of the line (on the left side) is the solution of this inequality.

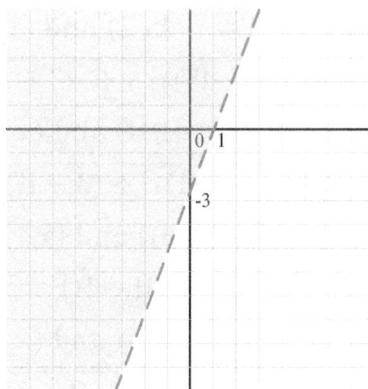

15) $y > -2x + 1$

$x = 0 \rightarrow y = 0 + 1 = 1 \rightarrow (0, 1)$

$y = 0 \rightarrow 0 = -2x + 1 \rightarrow -2x = -1 \rightarrow x = \dfrac{-1}{-2}$

$\qquad = 0.5 \rightarrow (0.5, 0)$

The easiest point to test is the origin: $(0, 0)$

$(0,0) \rightarrow y > -2x + 1 \rightarrow 0 > -2(0) + 1 \rightarrow 0 > 1$

This is incorrect! 0 is less than 1. So, this part of the line (on the right side) is the solution of this inequality.

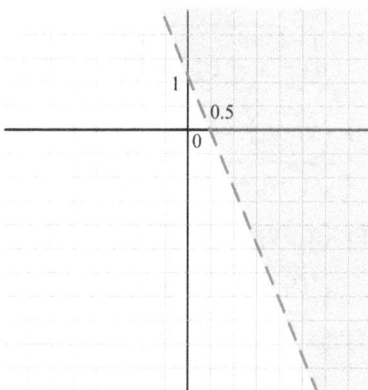

16) $M = \left(\dfrac{x_1 + x_2}{2}, \dfrac{y_1 + y_2}{2}\right) \rightarrow (x_1, y_1) = (4, 1)$ and $(x_2, y_2) = (2, 3) \rightarrow M = \left(\dfrac{4+2}{2}, \dfrac{1+3}{2}\right) \rightarrow$
$\left(\dfrac{6}{2}, \dfrac{4}{2}\right) \rightarrow M(3, 2)$

17) $(x_1, y_1) = (3, 6)$ and $(x_2, y_2) = (5, 4) \rightarrow M = \left(\dfrac{3+5}{2}, \dfrac{6+4}{2}\right) \rightarrow \left(\dfrac{8}{2}, \dfrac{10}{2}\right) \rightarrow M(4, 5)$

18) $(x_1, y_1) = (7, 1)$ and $(x_2, y_2) = (1, 3) \rightarrow M = \left(\dfrac{7+1}{2}, \dfrac{1+3}{2}\right) \rightarrow \left(\dfrac{8}{2}, \dfrac{4}{2}\right) \rightarrow M(4, 2)$

19) $(x_1, y_1) = (2, 8)$ and $(x_2, y_2) = (2, 10) \rightarrow M = \left(\dfrac{2+2}{2}, \dfrac{8+10}{2}\right) \rightarrow \left(\dfrac{4}{2}, \dfrac{18}{2}\right) \rightarrow M(2, 9)$

20) $(x_1, y_1) = (3, -2)$ and $(x_2, y_2) = (-1, 6) \rightarrow M = \left(\dfrac{3-1}{2}, \dfrac{-2+6}{2}\right) \rightarrow \left(\dfrac{2}{2}, \dfrac{4}{2}\right) \rightarrow M(1, 2)$

21) $(x_1, y_1) = (-1, -3)$ and $(x_2, y_2) = (1, 5) \rightarrow M = \left(\dfrac{-1+1}{2}, \dfrac{-3+5}{2}\right) \rightarrow \left(\dfrac{0}{2}, \dfrac{2}{2}\right) \rightarrow M(0, 1)$

22) $(x_1, y_1) = (1, 4)$ and $(x_2, y_2) = (-7, 6) \rightarrow M = \left(\dfrac{1-7}{2}, \dfrac{4+6}{2}\right) \rightarrow \left(\dfrac{-6}{2}, \dfrac{10}{2}\right) \rightarrow M(-3, 5)$

23) $(x_1, y_1) = (-3, 5)$ and $(x_2, y_2) = (7, -9) \rightarrow M = \left(\dfrac{-3+7}{2}, \dfrac{5-9}{2}\right) \rightarrow \left(\dfrac{4}{2}, \dfrac{-4}{2}\right) \rightarrow M(2, -2)$

24) $(x_1, y_1) = (-8, -1)$ and $(x_2, y_2) = (-4, 2) \rightarrow d = \sqrt{(x_2 - x_1)^2 + (y_2 - y_1)^2} =$
$\sqrt{(-4 - (-8))^2 + (2 - (-1))^2} = \sqrt{(4)^2 + (3)^2} = \sqrt{16 + 9} = \sqrt{25} = 5$

25) $(x_1, y_1) = (-15, 2)$ and $(x_2, y_2) = (5, -13) \rightarrow d = \sqrt{(x_2 - x_1)^2 + (y_2 - y_1)^2} = $
$\sqrt{(5 - (-15))^2 + (-13 - 2)^2} = \sqrt{(20)^2 + (-15)^2} = \sqrt{400 + 225} = \sqrt{625} = 25$

26) $(x_1, y_1) = (-1, 11)$ and $(x_2, y_2) = (-7, 3) \rightarrow d = \sqrt{(x_2 - x_1)^2 + (y_2 - y_1)^2} = $
$\sqrt{(-7 - (-1))^2 + (3 - 11)^2} = \sqrt{(-6)^2 + (-8)^2} = \sqrt{36 + 64} = \sqrt{100} = 10$

27) $(x_1, y_1) = (0, 11)$ and $(x_2, y_2) = (9, 11) \rightarrow d = \sqrt{(x_2 - x_1)^2 + (y_2 - y_1)^2} = $
$\sqrt{(9 - 0)^2 + (11 - 11)^2} = \sqrt{(9)^2 + (0)^2} = \sqrt{81} = 9$

28) $(x_1, y_1) = (-2, 4)$ and $(x_2, y_2) = (3, -8) \rightarrow d = \sqrt{(x_2 - x_1)^2 + (y_2 - y_1)^2} = $
$\sqrt{(3 - (-2))^2 + (-8 - 4)^2} = \sqrt{(5)^2 + (-12)^2} = \sqrt{25 + 144} = \sqrt{169} = 13$

29) $(x_1, y_1) = (6, -7)$ and $(x_2, y_2) = (-9, 1) \rightarrow d = \sqrt{(x_2 - x_1)^2 + (y_2 - y_1)^2} = $
$\sqrt{(-9 - 6)^2 + (1 - (-7))^2} = \sqrt{(-15)^2 + (8)^2} = \sqrt{225 + 64} = \sqrt{289} = 17$

30) $(x_1, y_1) = (8, -4)$ and $(x_2, y_2) = (-4, -20) \rightarrow d = \sqrt{(x_2 - x_1)^2 + (y_2 - y_1)^2} = $
$\sqrt{(-4 - 8)^2 + (-20 - (-4))^2} = \sqrt{(-12)^2 + (-16)^2} = \sqrt{144 + 256} = \sqrt{400} = 20$

31) $(x_1, y_1) = (5, 1)$ and $(x_2, y_2) = (9, -2) \rightarrow d = \sqrt{(x_2 - x_1)^2 + (y_2 - y_1)^2} = $
$\sqrt{(9 - 5)^2 + (-2 - 1)^2} = \sqrt{(4)^2 + (-3)^2} = \sqrt{16 + 9} = \sqrt{25} = 5$

32) $(x_1, y_1) = (-8, -17)$ and $(x_2, y_2) = (2, 7) \rightarrow d = \sqrt{(x_2 - x_1)^2 + (y_2 - y_1)^2} = $
$\sqrt{(2 - (-8))^2 + (7 - (-17))^2} = \sqrt{(10)^2 + (-24)^2} = \sqrt{100 + 576} = \sqrt{676} = 26$

33) $(x_1, y_1) = (18, 21)$ and $(x_2, y_2) = (-12, 5) \rightarrow d = \sqrt{(x_2 - x_1)^2 + (y_2 - y_1)^2} = $
$\sqrt{(-12 - 18)^2 + (5 - 21)^2} = \sqrt{(-30)^2 + (-16)^2} = \sqrt{900 + 256} = \sqrt{1,156} = 34$

DAY 6 Polynomials

Math topics that you'll learn in this chapter:

1. Simplifying Polynomials
2. Adding and Subtracting Polynomials
3. Multiplying Monomials
4. Multiplying and Dividing Monomials
5. Multiplying a Polynomial and a Monomial
6. Multiplying Binomials
7. Factoring Trinomials

63

Simplifying Polynomials

☆ To simplify Polynomials, find "like" terms. (they have same variables with same power).

☆ Use "FOIL". (First–Out–In–Last) for binomials:

$$(x + a)(x + b) = x^2 + (b + a)x + ab$$

☆ Add or Subtract "like" terms using order of operation.

Examples:

Example 1. Simplify this expression. $2x(3x - 4) - 6x =$

Solution: Use Distributive Property: $2x(3x - 4) = 6x^2 - 8x$

Now, combine like terms: $2x(3x - 4) - 6x = 6x^2 - 8x - 6x = 6x^2 - 14x$

Example 2. Simplify this expression. $(x + 5)(x + 7) =$

Solution: First, apply the FOIL method: $(a + b)(c + d) = ac + ad + bc + bd$

$(x + 5)(x + 7) = x^2 + 5x + 7x + 35$

Now combine like terms: $x^2 + 5x + 7x + 35 = x^2 + 12x + 35$

Example 3. Simplify this expression. $3x(-4x + 5) + 2x^2 - 5x =$

Solution: Use Distributive Property: $3x(-4x + 5) = -12x^2 + 15x$

Then: $3x(-4x + 5) + 2x^2 - 5x = -12x^2 + 15x + 2x^2 - 5x$

Now combine like terms: $-12x^2 + 2x^2 = -10x^2$, and $15x - 5x = 10x$

The simplified form of the expression: $-12x^2 + 15x + 2x^2 - 5x = -10x^2 + 10x$

Adding and Subtracting Polynomials

☆ Adding polynomials is just a matter of combining like terms, with some order of operations considerations thrown in.

☆ Be careful with the minus signs, and don't confuse addition and multiplication!

☆ For subtracting polynomials, sometimes you need to use the Distributive Property: $a(b + c) = ab + ac$, $a(b - c) = ab - ac$

Examples:

Example 1. Simplify the expressions. $(-5x^2 + 2x^3) - (-4x^3 + 2x^2) =$

Solution: First, use Distributive Property:

$-(-4x^3 + 2x^2) = 4x^3 - 2x^2$

$\rightarrow (-5x^2 + 2x^3) - (-4x^3 + 2x^2) = -5x^2 + 2x^3 + 4x^3 - 2x^2$

Now combine like terms: $2x^3 + 4x^3 = 6x^3$ and $-5x^2 - 2x^2 = -7x^2$

Then: $(x^2 - 2x^3) - (x^3 - 3x^2) = 6x^3 - 7x^2$

Example 2. Add expressions. $(2x^3 + 8) + (-x^3 + 4x^2) =$

Solution: Remove parentheses:

$$(2x^3 + 8) + (-x^3 + 4x^2) = 2x^3 + 8 - x^3 + 4x^2$$

Now combine like terms: $2x^3 + 8 - x^3 + 4x^2 = x^3 + 4x^2 + 8$

Example 3. Simplify the expressions. $(x^2 + 7x^3) - (11x^2 - 4x^3) =$

Solution: First, use Distributive Property: $-(11x^2 - 4x^3) = -11x^2 + 4x^3 \rightarrow$

$$(x^2 + 7x^3) - (11x^2 - 4x^3) = x^2 + 7x^3 - 11x^2 + 4x^3$$

Now combine like terms and write in standard form:

$x^2 + 7x^3 - 11x^2 + 4x^3 = 11x^3 - 10x^2$

Multiplying Monomials

✭ A monomial is a polynomial with just one term: Examples: $2x$ or $7y^2$.

✭ When you multiply monomials, first multiply the coefficients (a number placed before and multiplying the variable) and then multiply the variables using multiplication property of exponents.

$$x^a \times x^b = x^{a+b}$$

Examples:

Example 1. Multiply expressions. $3x^4y^5 \times 6x^2y^3$

Solution: Find the same variables and use multiplication property of exponents:
$x^a \times x^b = x^{a+b}$
$x^4 \times x^2 = x^{4+2} = x^6$ and $y^5 \times y^3 = y^{5+3} = y^8$
Then, multiply coefficients and variables: $3x^4y^5 \times 6x^2y^3 = 18x^6y^8$

Example 2. Multiply expressions. $5a^5b^9 \times 3a^2b^8 =$

Solution: Use the multiplication property of exponents: $x^a \times x^b = x^{a+b}$
$a^5 \times a^2 = a^{5+2} = a^7$ and $b^9 \times b^8 = b^{9+8} = b^{17}$
Then: $5a^5b^9 \times 3a^2b^8 = 15a^7b^{17}$

Example 3. Multiply. $6x^3y^2z^4 \times 2x^2y^8z^6$

Solution: Use the multiplication property of exponents: $x^a \times x^b = x^{a+b}$
$x^3 \times x^2 = x^{3+2} = x^5$, $y^2 \times y^8 = y^{2+8} = y^{10}$ and $z^4 \times z^6 = z^{4+6} = z^{10}$
Then: $6x^3y^2z^4 \times 2x^2y^8z^6 = 12x^5y^{10}z^{10}$

Example 4. Simplify. $(2a^3b^6)(-5a^7b^{12}) =$

Solution: Use the multiplication property of exponents: $x^a \times x^b = x^{a+b}$
$a^3 \times a^7 = a^{3+7} = a^{10}$ and $b^6 \times b^{12} = b^{6+12} = b^{18}$
Then: $(2a^3b^6)(-5a^7b^{12}) = -10a^{10}b^{18}$

Multiplying and Dividing Monomials

☆ When you divide or multiply two monomials, you need to divide or multiply their coefficients and then divide or multiply their variables.

☆ In case of exponents with the same base, for Division, subtract their powers, for Multiplication, add their powers.

☆ Exponent's Multiplication and Division rules:

$$x^a \times x^b = x^{a+b}, \qquad \frac{x^a}{x^b} = x^{a-b}$$

Examples:

Example 1. Multiply expressions. $(5x^4)(3x^9) =$

Solution: Use multiplication property of exponents:
$x^a \times x^b = x^{a+b} \rightarrow x^4 \times x^9 = x^{13}$
Then: $(5x^4)(3x^9) = 15x^{13}$

Example 2. Divide expressions. $\frac{18x^3y^6}{9x^2y^4} =$

Solution: Use division property of exponents:
$\frac{x^a}{x^b} = x^{a-b} \rightarrow \frac{x^3}{x^2} = x^{3-2} = x^1$ and $\frac{y^6}{y^4} = y^{6-4} = y^2$
Then: $\frac{18x^3y^6}{9x^2y^4} = 2xy^2$

Example 3. Divide expressions. $\frac{51a^4b^{11}}{3a^2b^5}$

Solution: Use division property of exponents:
$\frac{x^a}{x^b} = x^{a-b} \rightarrow \frac{a^4}{a^2} = a^{4-2} = a^2$ and $\frac{b^{11}}{b^5} = b^{11-5} = b^6$
Then. $\frac{51a^4b^{11}}{3a^2b^5} = 17a^2b^6$

Multiplying a Polynomial and a Monomial

☆ When multiplying monomials, use the product rule for exponents.

$$x^a \times x^b = x^{a+b}$$

☆ When multiplying a monomial by a polynomial, use the distributive property.

$$a \times (b + c) = a \times b + a \times c = ab + ac$$
$$a \times (b - c) = a \times b - a \times c = ab - ac$$

Examples:

Example 1. Multiply expressions. $5x(4x + 7)$

Solution: Use Distributive Property:

$5x(4x + 7) = (5x \times 4x) + (5x \times 7) = 20x^2 + 35x$

Example 2. Multiply expressions. $y(2x^2 + 3y^2)$

Solution: Use Distributive Property:

$y(2x^2 + 3y^2) = y \times 2x^2 + y \times 3y^2 = 2x^2y + 3y^3$

Example 3. Multiply. $-2x(-x^2 + 3x + 6)$

Solution: Use Distributive Property:

$-2x(-x^2 + 3x + 6) = (-2x)(-x^2) + (-2x) \times (3x) + (-2x) \times (6) =$
Now simplify:

$(-2x)(-x^2) + (-2x) \times (3x) + (-2x) \times (6) = 2x^3 - 6x^2 - 12x$

Multiplying Binomials

☆ A binomial is a polynomial that is the sum or the difference of two terms, each of which is a monomial.

☆ To multiply two binomials, use the "FOIL" method. (First–Out–In–Last)

$$(x + a)(x + b) = x \times x + x \times b + a \times x + a \times b = x^2 + bx + ax + ab$$

Examples:

Example 1. Multiply Binomials. $(x - 5)(x + 4) =$

Solution: Use "FOIL". (First–Out–In–Last):

$(x - 5)(x + 4) = x^2 - 5x + 4x - 20$

Then combine like terms: $x^2 - 5x + 4x - 20 = x^2 - x - 20$

Example 2. Multiply. $(x + 3)(x + 6) =$

Solution: Use "FOIL". (First–Out–In–Last):

$(x + 3)(x + 6) = x^2 + 3x + 6x + 18$

Then simplify: $x^2 + 3x + 6x + 18 = x^2 + 9x + 18$

Example 3. Multiply. $(x - 8)(x + 4) =$

Solution: Use "FOIL". (First–Out–In–Last):

$(x - 8)(x + 4) = x^2 - 8x + 4x - 32$

Then simplify: $x^2 - 8x + 4x - 32 = x^2 - 4x - 32$

Example 4. Multiply Binomials. $(x - 6)(x - 3) =$

Solution: Use "FOIL". (First–Out–In–Last):

$(x - 6)(x - 3) = x^2 - 6x - 3x + 18$

Then combine like terms: $x^2 - 6x - 3x + 18 = x^2 - 9x + 18$

Factoring Trinomials

To factor trinomials, you can use following methods:

☆ "FOIL": $(x + a)(x + b) = x^2 + (b + a)x + ab$

☆ "Difference of Squares":

$$a^2 - b^2 = (a + b)(a - b)$$
$$a^2 + 2ab + b^2 = (a + b)(a + b)$$
$$a^2 - 2ab + b^2 = (a - b)(a - b)$$

☆ "Reverse FOIL": $x^2 + (b + a)x + ab = (x + a)(x + b)$

Examples:

Example 1. Factor this trinomial. $x^2 - 3x - 18$

Solution: Break the expression into groups. You need to find two numbers that their product is -18 and their sum is -3. (remember "Reverse FOIL": $x^2 + (b + a)x + ab = (x + a)(x + b)$). Those two numbers are 3 and -6. Then:
$$x^2 - 3x - 18 = (x^2 + 3x) + (-6x - 18)$$
Now factor out x from $x^2 + 3x : x(x + 3)$, and factor out -6 from
$-6x - 18: -6(x + 3)$; Then: $(x^2 + 3x) + (-6x - 18) = x(x + 3) - 6(x + 3)$
Now factor out like term: $(x + 3)$. Then: $(x + 3)(x - 6)$

Example 2. Factor this trinomial. $2x^2 - 4x - 48$

Solution: Break the expression into groups: $(2x^2 + 8x) + (-12x - 48)$
Now factor out $2x$ from $2x^2 + 8x : 2x(x + 4)$, and factor out -12 from
$- 12x - 48: -12(x + 4)$; Then: $2x(x + 4) - 12(x + 4)$, now factor out like term:
$(x + 4) \rightarrow 2x(x + 4) - 12(x + 4) = (x + 4)(2x - 12)$

Day 6: Practices

✎ Simplify each polynomial.

1) $4(3x + 2) =$

2) $7(6x - 3) =$

3) $x(4x + 5) + 6x =$

4) $2x(x - 4) + 8x =$

5) $3(5x + 3) - 9x =$

6) $x(6x - 5) - 4x^2 + 11 =$

7) $-x^2 + 7 + 3x(x + 2) =$

8) $6x^2 - 7 + 3x(5x - 7) =$

✎ Add or subtract polynomials.

9) $(x^2 + 5) + (3x^2 - 2) =$

10) $(4x^2 - 5x) - (x^2 + 7x) =$

11) $(6x^3 - 2x^2) + (3x^3 - 7x^2) =$

12) $(6x^3 - 7x) - (9x^3 - 3x) =$

13) $(9x^3 + 5x^2) + (12x^2 - 7) =$

14) $(5x^3 - 8) - (2x^3 - 6x^2) =$

15) $(10x^3 + 4x) - (7x^3 - 5x) =$

16) $(12x^3 - 7x) - (3x^3 + 9x) =$

✎ Find the products. (Multiplying Monomials)

17) $5x^3 \times 6x^5 =$

18) $3x^4 \times 4x^3 =$

19) $-7a^3b \times 3a^2b^5 =$

20) $-5x^2y^3z \times 6x^6y^4z^5 =$

21) $-2a^3bc \times (-4a^8b^7) =$

22) $9u^6t^5 \times (-2u^2t) =$

23) $14x^2z \times 2x^6y^8z =$

24) $-12x^7y^6z \times 3xy^8 =$

25) $-9a^2b^3c \times 3a^7b^6 =$

26) $-11x^9y^7 \times (-6x^4y^3) =$

✍ **Simplify each expression. (Multiplying and Dividing Monomials)**

27) $(4x^3y^4)(2x^4y^3) =$

28) $(7x^2y^5)(3x^3y^6) =$

29) $(5x^9y^6)(8x^6y^9) =$

30) $(13a^4b^7)(2a^6b^9) =$

31) $\frac{54x^6y^3}{9x^4y} =$

32) $\frac{28x^5y^7}{4x^3y^4} =$

33) $\frac{32x^{17}y^{12}}{8x^{13}y^9} =$

34) $\frac{40x^7y^{19}}{5x^2y^{14}} =$

✍ **Find each product. (Multiplying a Polynomial and a Monomial)**

35) $6(4x - 2y) =$

36) $4x(5x + y) =$

37) $8x(x - 2y) =$

38) $x(3x^2 + 4x - 6) =$

39) $4x(2x^2 + 7x + 4) =$

40) $8x(3x^2 - 7x - 3) =$

✍ **Find each product. (Multiplying Binomials)**

41) $(x - 4)(x + 5) =$

42) $(x - 3)(x + 3) =$

43) $(x + 8)(x + 7) =$

44) $(x - 5)(x + 9) =$

45) $(2x + 4)(x - 6) =$

46) $(2x - 11)(x + 6) =$

✍ **Factor each trinomial.**

47) $x^2 + 2x - 15 =$

48) $x^2 - x - 42 =$

49) $x^2 - 14x + 49 =$

50) $x^2 - 7x - 60 =$

51) $2x^2 + 6x - 20 =$

52) $3x^2 + 13x - 10 =$

Day 6: Answers

1) $4(3x + 2) = (4 \times 3x) + (4 \times 2) = 12x + 8$

2) $7(6x - 3) = (7 \times 6x) - (7 \times 3) = 42x - 21$

3) $x(4x + 5) + 6x = (x \times 4x) + (x \times 5) + 6x = 4x^2 + 5x + 6x = 4x^2 + 11x$

4) $2x(x - 4) + 8x = (2x \times x) + (2x \times (-4)) + 8x = 2x^2 - 8x + 8x = 2x^2$

5) $3(5x + 3) - 9x = (3 \times 5x) + (3 \times 3) - 9x = 15x - 9x + 9 = 6x + 9$

6) $x(6x - 5) - 4x^2 + 11 = (x \times 6x) + (x \times (-5)) - 4x^2 + 11 =$
$6x^2 - 4x^2 - 5x + 11 = 2x^2 - 5x + 11$

7) $-x^2 + 7 + 3x(x + 2) = -x^2 + 7 + (3x \times x) + (3x \times 2) = -x^2 + 7 + 3x^2 + 6x =$
$2x^2 + 6x + 7$

8) $6x^2 - 7 + 3x(5x - 7) = 6x^2 - 7 + (3x \times 5x) + (3x \times (-7)) =$
$6x^2 - 7 + 15x^2 - 21x = 21x^2 - 21x - 7$

9) $(x^2 + 5) + (3x^2 - 2) = x^2 + 3x^2 + 5 - 2 = 4x^2 + 3$

10) $(4x^2 - 5x) - (x^2 + 7x) = 4x^2 - x^2 - 5x - 7x = 3x^2 - 12x$

11) $(6x^3 - 2x^2) + (3x^3 - 7x^2) = 6x^3 + 3x^3 - 2x^2 - 7x^2 = 9x^3 - 9x^2$

12) $(6x^3 - 7x) - (9x^3 - 3x) = 6x^3 - 9x^3 - 7x + 3x = -3x^3 - 4x$

13) $(9x^3 + 5x^2) + (12x^2 - 7) = 9x^3 + 5x^2 + 12x^2 - 7 = 9x^3 + 17x^2 - 7$

14) $(5x^3 - 8) - (2x^3 - 6x^2) = 5x^3 - 2x^3 + 6x^2 - 8 = 3x^3 + 6x^2 - 8$

15) $(10x^3 + 4x) - (7x^3 - 5x) = 10x^3 - 7x^3 + 4x + 5x = 3x^3 + 9x$

16) $(12x^3 - 7x) - (3x^3 + 9x) = 12x^3 - 3x^3 - 7x - 9x = 9x^3 - 16x$

17) $5x^3 \times 6x^5 \rightarrow 5 \times 6 = 30, \ x^3 \times x^5 = x^{3+5} = x^8 \rightarrow 5x^3 \times 6x^5 = 30x^8$

18) $3x^4 \times 4x^3 \rightarrow 3 \times 4 = 12, \ x^4 \times x^3 = x^{4+3} = x^7 \rightarrow 3x^4 \times 4x^3 = 12x^7$

19) $-7a^3b \times 3a^2b^5 \rightarrow -7 \times 3 = -21, \ a^3 \times a^2 = a^{3+2} = a^5, \ b \times b^5 = b^{1+5} = b^6 \rightarrow$
$-7a^3b \times 3a^2b^5 = -21a^5b^6$

20) $-5x^2y^3z \times 6x^6y^4z^5 \rightarrow -5 \times 6 = -30, \ x^2 \times x^6 = x^{2+6} = x^8, y^3 \times y^4 = y^{3+4} = y^7,$
$z \times z^5 = z^{1+5} = z^6 \rightarrow -5x^2y^3z \times 6x^6y^4z^5 = -30x^8y^7z^6$

21) $-2a^3bc \times (-4a^8b^7) \to -2 \times (-4) = 8, \; a^3 \times a^8 = a^{3+8} = a^{11}, b \times b^7 = b^{1+7} = b^8 \to -2a^3bc \times (-4a^8b^7) = 8a^{11}b^8c$

22) $9u^6t^5 \times (-2u^2t) \to 9 \times (-2) = -18, \; u^6 \times u^2 = u^{6+2} = u^8, t^5 \times t^1 = t^{5+1} = t^6 \to 9u^6t^5 \times (-2u^2t) = -18u^8t^6$

23) $14x^2z \times 2x^6y^8z \to 14 \times 2 = 28, x^2 \times x^6 = x^{2+6} = x^8, z \times z = z^{1+1} = z^2 \to 14x^2z \times 2x^6y^8z = 28x^8y^8z^2$

24) $-12x^7y^6z \times 3xy^8 \to -12 \times 3 = -36, x^7 \times x = x^{1+7} = x^8, \; y^6 \times y^8 = y^{6+8} = y^{14} \to -12x^7y^6z \times 3xy^8 = -36x^8y^{14}z$

25) $-9a^2b^3c \times 3a^7b^6 \to -9 \times 3 = -27, a^2 \times a^7 = a^{2+7} = a^9, \; b^3 \times b^6 = b^{3+6} = b^9 \to -9a^2b^3c \times 3a^7b^6 = -27a^9b^9c$

26) $-11x^9y^7 \times (-6x^4y^3) \to -11 \times (-6) = 66, x^9 \times x^4 = x^{9+4} = x^{13}, y^7 \times y^3 = y^{7+3} = y^{10} \to -11x^9y^7 \times (-6x^4y^3) = 66x^{13}y^{10}$

27) $(4x^3y^4)(2x^4y^3) \to 4 \times 2 = 8, x^3 \times x^4 = x^{3+4} = x^7, \; y^4 \times y^3 = y^{4+3} = y^7 \to (4x^3y^4)(2x^4y^3) = 8x^7y^7$

28) $(7x^2y^5)(3x^3y^6) \to 7 \times 3 = 21, x^2 \times x^3 = x^{2+3} = x^5, \; y^5 \times y^6 = y^{5+6} = y^{11} \to (7x^2y^5)(3x^3y^6) = 21x^5y^{11}$

29) $(5x^9y^6)(8x^6y^9) \to 5 \times 8 = 40, x^9 \times x^6 = x^{9+6} = x^{15}, \; y^6 \times y^9 = y^{6+9} = y^{15} \to (5x^9y^6)(8x^6y^9) = 40x^{15}y^{15}$

30) $(13a^4b^7)(2a^6b^9) \to 13 \times 2 = 26, a^4 \times a^6 = a^{4+6} = a^{10}, \; b^7 \times b^9 = b^{7+9} = b^{16} \to (13a^4b^7)(2a^6b^9) = 26a^{10}b^{16}$

31) $\frac{54x^6y^3}{9x^4y} \to \frac{54}{9} = 6, \; \frac{x^6}{x^4} = x^{6-4} = x^2, \; \frac{y^3}{y} = y^{3-1} = y^2 \to \frac{54x^6y^3}{9x^4y} = 6x^2y^2$

32) $\frac{28x^5y^7}{4x^3y^4} \to \frac{28}{4} = 7, \; \frac{x^5}{x^3} = x^{5-3} = x^2, \; \frac{y^7}{y^4} = y^{7-4} = y^3 \to \frac{28x^5y^7}{4x^3y^4} = 7x^2y^3$

33) $\frac{32x^{17}y^{12}}{8x^{13}y^9} \to \frac{32}{8} = 4, \; \frac{x^{17}}{x^{13}} = x^{17-13} = x^4, \; \frac{y^{12}}{y^9} = y^{12-9} = y^3 \to \frac{32x^{17}y^{12}}{8x^{13}y^9} = 4x^4y^3$

34) $\frac{40x^7y^{19}}{5x^2y^{14}} = \to \frac{40}{5} = 8, \; \frac{x^7}{x^2} = x^{7-2} = x^5, \; \frac{y^{19}}{y^{14}} = y^{19-14} = y^5 \to \frac{40x^7y^{19}}{5x^2y^{14}} = 8x^5y^5$

35) $6(4x - 2y) = (6 \times 4x) - (6 \times 2y) = 24x - 12y$

36) $4x(5x + y) = (4x \times 5x) + (4x \times y) = 20x^2 + 4xy$

37) $8x(x - 2y) = (8x \times x) - (8x \times 2y) = 8x^2 - 16xy$

38) $x(3x^2 + 4x - 6) = (x \times 3x^2) + (x \times 4x) + (x \times (-6)) = 3x^3 + 4x^2 - 6x$

39) $4x(2x^2 + 7x + 4) = (4x \times 2x^2) + (4x \times 7x) + (4x \times 4) = 8x^3 + 28x^2 + 16x$

40) $8x(3x^2 - 7x - 3) = (8x \times 3x^2) + (8x \times (-7x)) + (8x \times (-3)) = 24x^3 - 56x^2 - 24x$

41) $(x - 4)(x + 5) = (x \times x) + (x \times 5) + (-4 \times x) + (-4 \times 5) = x^2 + 5x - 4x - 20 =$
$x^2 + x - 20$

42) $(x - 3)(x + 3) = (x \times x) + (x \times 3) + (-3 \times x) + (-3 \times 3) = x^2 + 3x - 3x - 9 = x^2 - 9$

43) $(x + 8)(x + 7) = (x \times x) + (x \times 7) + (8 \times x) + (8 \times 7) = x^2 + 7x + 8x + 56 =$
$x^2 + 15x + 56$

44) $(x - 5)(x + 9) = (x \times x) + (x \times 9) + (-5 \times x) + (-5 \times 9) = x^2 + 9x - 5x - 45 =$
$x^2 + 4x - 45$

45) $(2x + 4)(x - 6) = (2x \times x) + (2x \times (-6)) + (4 \times x) + (4 \times (-6)) =$
$2x^2 - 12x + 4x - 24 = 2x^2 - 8x - 24$

46) $(2x - 11)(x + 6) = (2x \times x) + (2x \times 6) + ((-11) \times x) + ((-11) \times 6) =$
$2x^2 + 12x - 11x - 66 = 2x^2 + x - 66$

47) $x^2 + 2x - 15 \rightarrow$(Use this rule: $x^2 + (b + a)x + ab = (x + a)(x + b)$). Then:
$x^2 + 2x - 15 = x^2 + (5 - 3)x + (5 \times (-3)) = (x + 5)(x - 3)$

48) $x^2 - x - 42 = x^2 + (-7 + 6)x + ((-7) \times 6) = (x - 7)(x + 6)$

49) $x^2 - 14x + 49 = x^2 + (-7 - 7)x + ((-7) \times (-7)) = (x - 7)(x - 7)$

50) $x^2 - 7x - 60x^2 = x^2 + (-12 + 5)x + ((-12) \times 5) = (x - 12)(x + 5)$

51) $2x^2 + 6x - 20 = 2x^2 + (10x - 4x) - 20 = (2x^2 + 10x) + (-4x - 20) =$
$2x(x + 5) - 4(x + 5) = (2x - 4)(x + 5)$

52) $3x^2 + 13x - 10 = (3x^2 + 15x) + (-2x - 10) = 3x(x + 5) - 2(x + 5) =$
$(3x - 2)(x + 5)$

DAY 7

Geometry, Solid Figures and Probability

Math topics that you'll learn in this chapter:

1. The Pythagorean Theorem
2. Complementary and Supplementary angles
3. Parallel lines and Transversals
4. Triangles
5. Special Right Triangles
6. Polygons
7. Circles
8. Trapezoids
9. Cubes
10. Rectangle Prisms
11. Cylinder
12. Mean, Median, Mode, and Range of the Given Data
13. Pie Graph
14. Probability Problems
15. Permutations and Combinations

The Pythagorean Theorem

☆ You can use the Pythagorean Theorem to find a missing side in a right triangle.

☆ In any right triangle: $a^2 + b^2 = c^2$

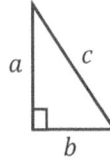

Examples:

Example 1. Right triangle ABC (not shown) has two legs of lengths 3 cm (AB) and 4 cm (AC). What is the length of the hypotenuse of the triangle (side BC)?

Solution: Use Pythagorean Theorem: $a^2 + b^2 = c^2$, $a = 3$ and $b = 4$

Then: $a^2 + b^2 = c^2 \rightarrow 3^2 + 4^2 = c^2 \rightarrow 9 + 16 = c^2 \rightarrow 25 = c^2 \rightarrow c = \sqrt{25} = 5$

The length of the hypotenuse is 5 cm.

Example 2. Find the hypotenuse of this triangle.

Solution: Use Pythagorean Theorem: $a^2 + b^2 = c^2$

Then: $a^2 + b^2 = c^2 \rightarrow 15^2 + 8^2 = c^2 \rightarrow 225 + 64 = c^2$

$c^2 = 289 \rightarrow c = \sqrt{289} = 17$

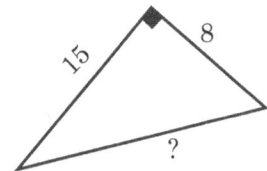

Example 3. Find the length of the missing side in this triangle.

Solution: Use Pythagorean Theorem: $a^2 + b^2 = c^2$

Then: $a^2 + b^2 = c^2 \rightarrow 20^2 + b^2 = 25^2 \rightarrow 400 + b^2 = 625 \rightarrow$

$b^2 = 625 - 400 \rightarrow b^2 = 225 \rightarrow b = \sqrt{225} = 15$

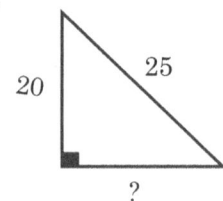

Complementary and Supplementary angles

☆ Two angles with a sum of 90 degrees are called complementary angles.

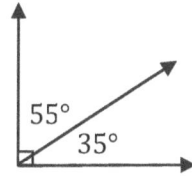

55°
35°

☆ Two angles with a sum of 180 degrees are Supplementary angles.

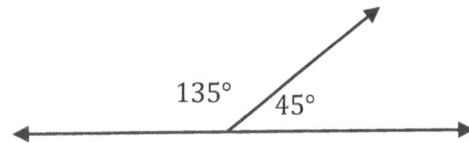

135° 45°

Examples:

Example 1. Find the missing angle.

Solution: Notice that the two angles form a right angle. This means that the angles are complementary, and their sum is 90. Then: $22° + x = 90° \rightarrow x = 90° - 22° = 68°$
The missing angle is 68 degrees. $x = 68°$

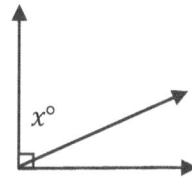

$x°$

Example 2. Angles Q and S are supplementary. What is the measure of angle Q if angle S is 45 degrees?

Solution: Q and S are supplementary $\rightarrow Q + S = 180 \rightarrow Q + 45 = 180 \rightarrow$
$$Q = 180 - 45 = 135°$$

Example 3. Angles x and y are complementary. What is the measure of angle x if angle y is 27 degrees?

Solution: Angles x and y are complementary $\rightarrow x + y = 90 \rightarrow x + 27 = 90 \rightarrow$
$$x = 90 - 27 = 63°$$

Parallel lines and Transversals

☆ When a line (transversal) intersects two parallel lines in the same plane, eight angles are formed. In the following diagram, a transversal intersects two parallel lines. Angles $1, 3, 5$ and 7 are congruent. Angles $2, 4, 6,$ and 8 are also congruent.

☆ In the following diagram, the following angles are supplementary angles (their sum is 180):

- ❖ Angles 1 and 8
- ❖ Angles 2 and 7
- ❖ Angles 3 and 6
- ❖ Angles 4 and 5

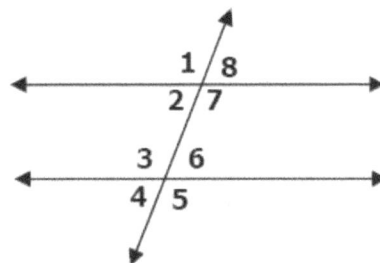

Example:

In the following diagram, two parallel lines are cut by a transversal. What is the value of x?

Solution: The two angles $2x + 4$ and $3x - 9$ are equivalent.

That is: $2x + 4 = 3x - 9$

Now, solve for x:

$2x + 4 + 9 = 3x - 9 + 9$

$\rightarrow 2x + 13 = 3x \rightarrow 2x + 13 - 2x = 3x - 2x \rightarrow$

$13 = x$

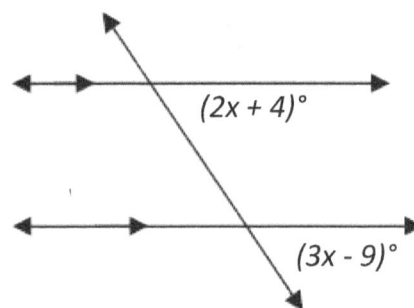

Triangles

☆ In any triangle, the sum of all angles is 180 degrees.

☆ Area of a triangle $= \frac{1}{2}$(base×height)

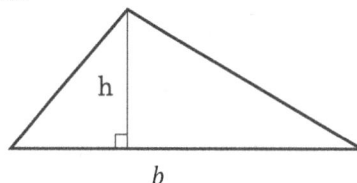

Examples:

Example 1. What is the area of this triangles?

Solution: Use the area formula:

Area$= \frac{1}{2}$(base×height)

base= 18 and height= 9, Then:

Area$= \frac{1}{2}(18 \times 9) = \frac{162}{2} = 81$

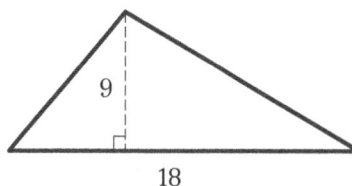

Example 2. What is the area of this triangles?

Solution: Use the area formula:

Area$= \frac{1}{2}$(base×height)

base= 18 and height= 7; Area$= \frac{1}{2}(18 \times 7) = \frac{126}{2} = 63$

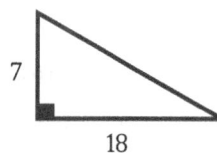

Example 3. What is the missing angle in this triangle?

Solution: In any triangle, the sum of all angles is
180 degrees. Let x be the missing angle.
Then: $63 + 84 + x = 180 \rightarrow 147 + x = 180 \rightarrow$
$x = 180 - 147 = 33°$
The missing angle is 33 degrees.

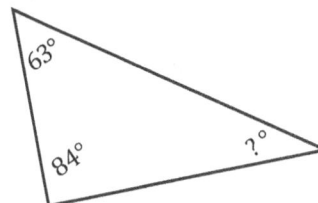

Special Right Triangles

☆ A special right triangle is a triangle whose sides are in a particular ratio. Two special right triangles are $45° - 45° - 90°$ and $30° - 60° - 90°$ triangles.

☆ In a special $45° - 45° - 90°$ triangle, the three angles are $45°$, $45°$ and $90°$. The lengths of the sides of this triangle are in the ratio of $1 : 1 : \sqrt{2}$.

☆ In a special triangle $30° - 60° - 90°$, the three angles are $30° - 60° - 90°$. The lengths of this triangle are in the ratio of $1 : \sqrt{3} : 2$.

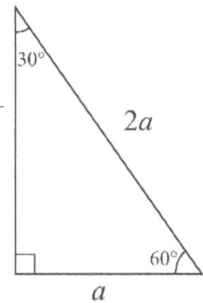

(triangle diagram with $30°$ angle at top, $60°$ angle at bottom right, right angle at bottom left, sides labeled $a\sqrt{3}$, $2a$, and a)

Examples:

Example 1. Find the length of the hypotenuse of a right triangle if the length of the other two sides are both 6 inches.

Solution: This is a right triangle with two equal sides. Therefore, it must be a $45° - 45° - 90°$ triangle. Two equivalent sides are 6 inches. The ratio of sides: $x : x : x\sqrt{2}$

The length of the hypotenuse is $6\sqrt{2}$ inches. $x : x : x\sqrt{2} \rightarrow 6 : 6 : 6\sqrt{2}$

Example 2. The length of the hypotenuse of a right triangle is 6 inches. What are the lengths of the other two sides if one angle of the triangle is $30°$?

Solution: The hypotenuse is 6 inches and the triangle is a $30° - 60° - 90°$ triangle. Then, one side of the triangle is 3 (it's half the side of the hypotenuse) and the other side is $3\sqrt{3}$. (it's the smallest side times $\sqrt{3}$)
$x : x\sqrt{3} : 2x \rightarrow x = 3 \rightarrow x : x\sqrt{3} : 2x = 3 : 3\sqrt{3} : 6$

Polygons

☆ The perimeter of a square = $4 \times side = 4s$

☆ The perimeter of a rectangle = $2(width + length)$

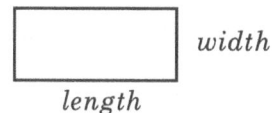

☆ The perimeter of trapezoid = $a + b + c + d$

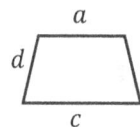

☆ The perimeter of a regular hexagon = $6a$

☆ The perimeter of a parallelogram = $2(l + w)$

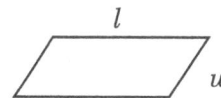

Examples:

Example 1. Find the perimeter of following regular hexagon.

Solution: Since the hexagon is regular, all sides are equal.
Then, the perimeter of the hexagon = $6 \times (one\ side)$
The perimeter of the hexagon = $6 \times (one\ side) = 6 \times 7 = 42\ m$

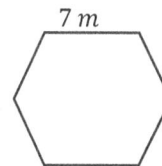

Example 2. Find the perimeter of following trapezoid.

Solution: The perimeter of a trapezoid = $a + b + c + d$
The perimeter of the trapezoid = $9 + 7 + 13 + 7 = 36\ ft$

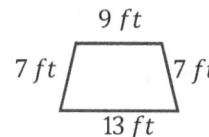

Circles

⭐ In a circle, variable r is usually used for the radius and d for diameter.

⭐ *Area of a circle* $= \pi r^2$ (π is about 3.14)

⭐ *Circumference of a circle* $= 2\pi r$

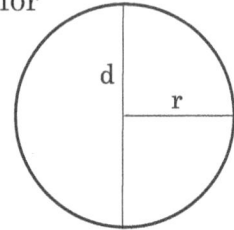

Examples:

Example 1. Find the area of this circle. ($\pi = 3.14$)

Solution:
Use area formula: $Area = \pi r^2$
$r = 5\ in \rightarrow Area = \pi(5)^2 = 25\pi$, $\pi = 3.14$
Then: $Area = 25 \times 3.14 = 78.5\ in^2$

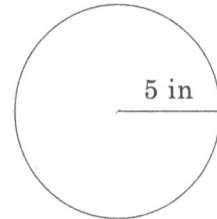

Example 2. Find the Circumference of this circle. ($\pi = 3.14$)

Solution:
Use Circumference formula: $Circumference = 2\pi r$
$r = 7\ cm \rightarrow Circumference = 2\pi(7) = 14\pi$
$\pi = 3.14$, Then: $Circumference = 14 \times 3.14 = 43.96\ cm$

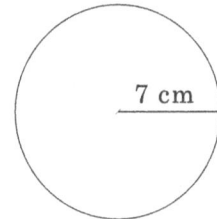

Example 3. Find the area of this circle.

Solution:
Use area formula: $Area = \pi r^2$
$r = 11\ in$, Then: $Area = \pi(11)^2 = 121\pi$, $\pi = 3.14$
$Area = 121 \times 3.14 = 379.94\ in^2$

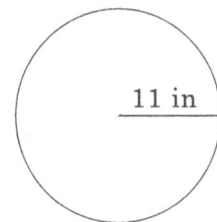

Trapezoids

☆ A quadrilateral with at least one pair of parallel sides is a trapezoid.

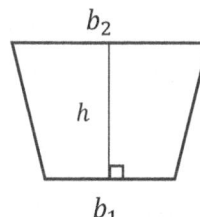

☆ Area of a trapezoid $= \frac{1}{2}h(b_1 + b_2)$

Examples:

Example 1. Calculate the area of this trapezoid.

Solution:

Use area formula: $A = \frac{1}{2}h(b_1 + b_2)$

$b_1 = 5\ cm$, $b_2 = 9\ cm$ and $h = 16\ cm$

Then: $A = \frac{1}{2}(16)(9 + 5) = 8(14) = 112\ cm^2$

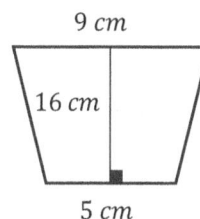

Example 2. Calculate the area of this trapezoid.

Solution:

Use area formula: $A = \frac{1}{2}h(b_1 + b_2)$

$b_1 = 8\ cm$, $b_2 = 16\ cm$ and $h = 12\ cm$

Then: $A = \frac{1}{2}(12)(8 + 16) = 144\ cm^2$

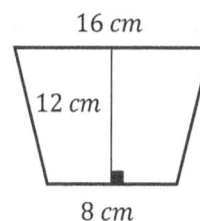

Cubes

☆ A cube is a three-dimensional solid object bounded by six square sides.

☆ Volume is the measure of the amount of space inside of a solid figure, like a cube, ball, cylinder or pyramid.

☆ The volume of a cube = $(one\ side)^3$

☆ The surface area of a cube = $6 \times (one\ side)^2$

Examples:

Example 1. Find the volume and surface area of this cube.

Solution: Use volume formula: $volume = (one\ side)^3$
Then: $volume = (one\ side)^3 = (5)^3 = 125\ cm^3$
Use surface area formula:
$surface\ area\ of\ a\ cube: 6(one\ side)^2 = 6(5)^2 = 6(25) = 150\ cm^2$

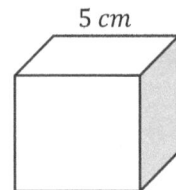

5 cm

Example 2. Find the volume and surface area of this cube.

Solution: Use volume formula: $volume = (one\ side)^3$
Then: $volume = (one\ side)^3 = (7)^3 = 343\ cm^3$
Use surface area formula:
$surface\ area\ of\ a\ cube: 6(one\ side)^2 = 6(7)^2 = 6(49) = 294\ cm^2$

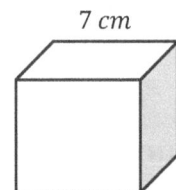

7 cm

Example 3. Find the volume and surface area of this cube.

Solution: Use volume formula: $volume = (one\ side)^3$
Then: $volume = (one\ side)^3 = (9)^3 = 729\ m^3$
Use surface area formula:
$surface\ area\ of\ a\ cube: 6(one\ side)^2 = 6(9)^2 = 6(81) = 486\ m^2$

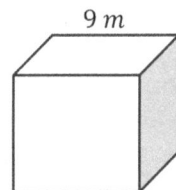

9 m

Rectangular Prisms

✫ A rectangular prism is a solid 3-dimensional object with six rectangular faces.

✫ The volume of a Rectangular prism = $Length \times Width \times Height$

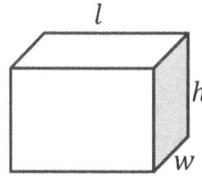

$Volume = l \times w \times h$

$Surface\ area = 2 \times (wh + lw + lh)$

Examples:

Example 1. Find the volume and surface area of this rectangular prism.

Solution: Use volume formula: $Volume = l \times w \times h$

Then: $Volume = 9 \times 7 \times 10 = 630\ m^3$

Use surface area formula: $Surface\ area = 2 \times (wh + lw + lh)$

Then: $Surface\ area = 2 \times ((7 \times 10) + (9 \times 7) + (9 \times 10))$

$$= 2 \times (70 + 63 + 90) = 2 \times (223) = 446\ m^2$$

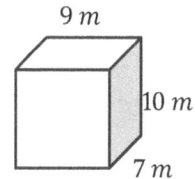

Example 2. Find the volume and surface area of this rectangular prism.

Solution: Use volume formula: $Volume = l \times w \times h$

Then: $Volume = 8 \times 5 \times 11 = 440\ m^3$

Use surface area formula: $Surface\ area = 2 \times (wh + lw + lh)$

Then: $Surface\ area = 2 \times ((5 \times 11) + (8 \times 5) + (8 \times 11))$

$$= 2 \times (55 + 40 + 88) = 2 \times (183) = 366\ m^2$$

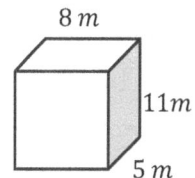

Cylinder

☆ A cylinder is a solid geometric figure with straight parallel sides and a circular or oval cross-section.

☆ *Volume of a Cylinder* $= \pi(radius)^2 \times height$, $\pi \approx 3.14$

☆ *Surface area of a cylinder* $= 2\pi r^2 + 2\pi rh$

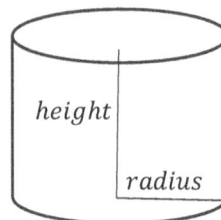

Examples:

Example 1. Find the volume and Surface area of the follow Cylinder.

Solution: Use volume formula:

$Volume = \pi(radius)^2 \times height$

Then: $Volume = \pi(6)^2 \times 12 = 36\pi \times 12 = 432\pi$

$\pi = 3.14$, then: $Volume = 432\pi = 432 \times 3.14 = 1,356.48 \, cm^3$

Use surface area formula: $Surface \, area = 2\pi r^2 + 2\pi rh$

Then: $2\pi(6)^2 + 2\pi(6)(12) = 2\pi(36) + 2\pi(72) = 72\pi + 144\pi = 216\pi$

$\pi = 3.14$, Then: $Surface \, area = 216 \times 3.14 = 678.24 \, cm^2$

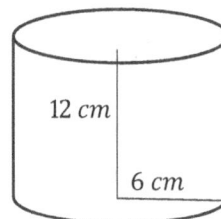

Example 2. Find the volume and Surface area of the follow Cylinder.

Solution: Use volume formula:

$Volume = \pi(radius)^2 \times height$

Then: $Volume = \pi(2)^2 \times 5 = 4\pi \times 5 = 20\pi$

$\pi = 3.14$, Then: $Volume = 20\pi = 62.8 \, cm^3$

Use surface area formula: $Surface \, area = 2\pi r^2 + 2\pi rh$

Then: $= 2\pi(2)^2 + 2\pi(2)(5) = 2\pi(4) + 2\pi(10) = 8\pi + 20\pi = 28\pi$

$\pi = 3.14$, then: $Surface \, area = 28 \times 3.14 = 87.92 \, cm^2$

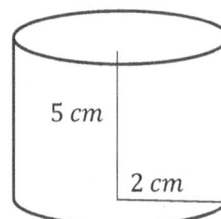

Mean, Median, Mode, and Range of the Given Data

☆ **Mean:** $\dfrac{sum\ of\ the\ data}{total\ number\ of\ data\ entires}$

☆ **Mode:** the value in the list that appears most often

☆ **Median:** is the middle number of a group of numbers arranged in order by size.

☆ **Range:** the difference of the largest value and smallest value in the list

Examples:

Example 1. What is the mode of these numbers? $4, 7, 8, 7, 8,\ 9,\ 8,\ 5$

Solution: Mode: the value in the list that appears most often.
Therefore, the mode is number 8. There are three number 8 in the data.

Example 2. What is the median of these numbers? $6, 11, 15, 10, 17, 20, 7$

Solution: Write the numbers in order: $6, 7, 10, 11, 15, 17, 20$
The median is the number in the middle. Therefore, the median is 11.

Example 3. What is the mean of these numbers? $8, 5, 3, 7, 6, 4, 9$

Solution: Mean: $\dfrac{sum\ of\ the\ data}{total\ number\ of\ data\ entires} = \dfrac{8+5+3+7+6+4+9}{7} = \dfrac{42}{7} = 6$

Example 4. What is the range in this list? $9, 2, 5, 10, 15, 22, 7$

Solution: Range is the difference of the largest value and smallest value in the list. The largest value is 22 and the smallest value is 2.
Then: $22 - 2 = 20$

Pie Graph

☆ A Pie Graph (Pie Chart) is a circle chart divided into sectors, each sector represents the relative size of each value.

☆ Pie charts represent a snapshot of how a group is broken down into smaller pieces.

Example:

A library has 650 books that include Mathematics, Physics, Chemistry, English and History. Use the following graph to answer the questions.

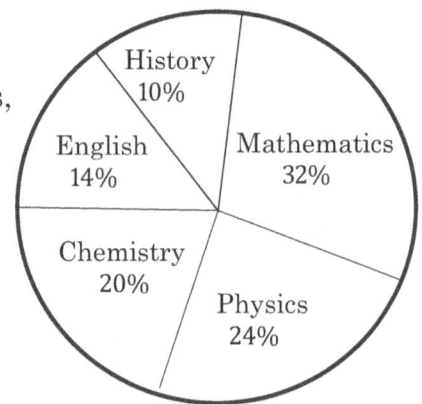

Example 1. What is the number of Mathematics books?

Solution: Number of total books = 650

Percent of Mathematics books = 32%

Then, the number of Mathematics books: 32% × 650 = 0.32 × 650 = 208

Example 2. What is the number of History books?

Solution: Number of total books = 650

Percent of History books = 10%

Then: 0.10 × 650 = 65

Example 3. What is the number of English books in the library?

Solution: Number of total books = 650

Percent of English books = 14%

Then: 0.14 × 650 = 91

Probability Problems

☆ Probability is the likelihood of something happening in the future. It is expressed as a number between zero (can never happen) to 1 (will always happen).

☆ Probability can be expressed as a fraction, a decimal, or a percent.

☆ Probability formula: $Probability = \frac{number\ of\ desired\ outcomes}{number\ of\ total\ outcomes}$

Examples:

Example 1. Anita's trick–or–treat bag contains 8 pieces of chocolate, 16 suckers, 22 pieces of gum and 20 pieces of licorice. If she randomly pulls a piece of candy from her bag, what is the probability of her pulling out a piece of gum?

Solution: Probability $= \frac{number\ of\ desired\ outcomes}{number\ of\ total\ outcomes}$

Probability of pulling out a piece of gum $= \frac{22}{8+16+22+20} = \frac{22}{66} = \frac{1}{3}$

Example 2. A bag contains 25 balls: five green, eight black, seven blue, a brown, a red and three white. If 24 balls are removed from the bag at random, what is the probability that a red ball has been removed?

Solution: If 24 balls are removed from the bag at random, there will be one ball in the bag. The probability of choosing a red ball is 1 out of 25. Therefore, the probability of not choosing a red ball is 24 out of 25 and the probability of having not a red ball after removing 24 balls is the same. The answer is: $\frac{24}{25}$

Permutations and Combinations

★ **Factorials** are products, indicated by an exclamation mark. For example, $4! = 4 \times 3 \times 2 \times 1$ (Remember that 0! is defined to be equal to 1)

★ **Permutations:** The number of ways to choose a sample of k elements from a set of n distinct objects where order does matter, and replacements are not allowed. For a permutation problem, use this formula:

$$nP_k = \frac{n!}{(n-k)!}$$

★ **Combination:** The number of ways to choose a sample of r elements from a set of n distinct objects where order does not matter, and replacements are not allowed. For a combination problem, use this formula:

$$nC_r = \frac{n!}{r! \, (n-r)!}$$

Examples:

Example 1. How many ways can the first and second place be awarded to 6 people?

Solution: Since the order matters, (the first and second place are different!) we need to use permutation formula where n is 6 and k is 2.
Then: $\frac{n!}{(n-k)!} = \frac{6!}{(6-2)!} = \frac{6!}{4!} = \frac{6 \times 5 \times 4!}{4!}$, remove 4! from both sides of the fraction. Then: $\frac{6 \times 5 \times 4!}{4!} = 6 \times 5 = 30$

Example 2. How many ways can we pick a team of 4 people from a group of 9?

Solution: Since the order doesn't matter, we need to use a combination formula where n is 9 and r is 4.
Then: $\frac{n!}{r! \, (n-r)!} = \frac{9!}{4! \, (9-4)!} = \frac{9!}{4! \, (5)!} = \frac{9 \times 8 \times 7 \times 6 \times 5!}{4! \, (5)!} = \frac{9 \times 8 \times 7 \times 6}{4 \times 3 \times 2 \times 1} = \frac{3,024}{24} = 126$

Day 7: Practices

✎ Find the missing side?

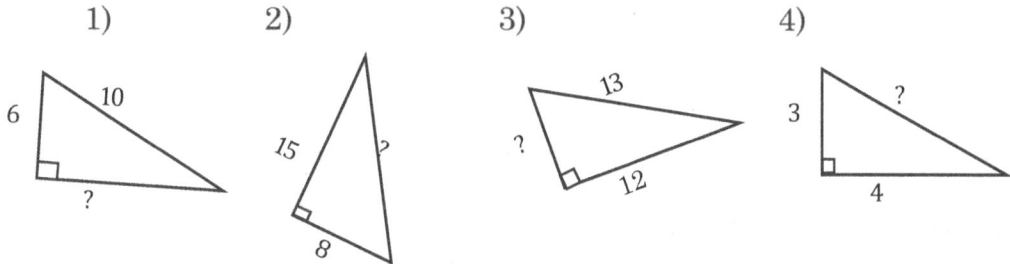

1)

6   10

?

2)

15   ?

8

3)

13

?   12

4)

3   ?

4

✎ Find the measure of the unknown angle in each triangle.

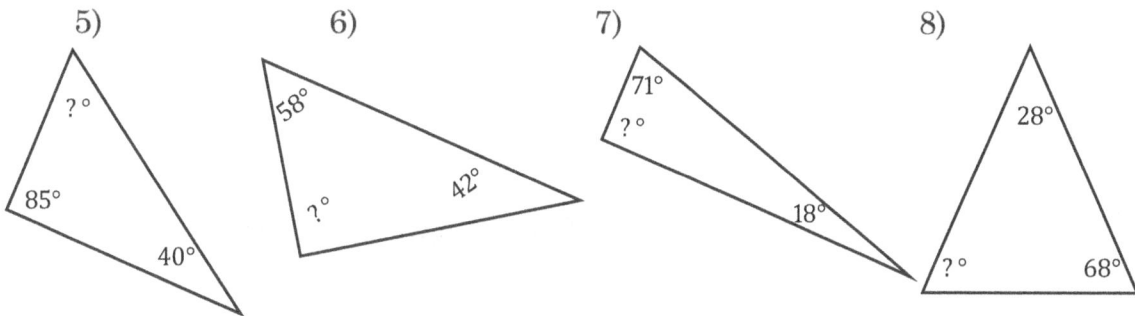

5)

?°

85°   40°

6)

58°

?°   42°

7)

71°
?°
18°

8)

28°

?°   68°

✎ Find the area of each triangle.

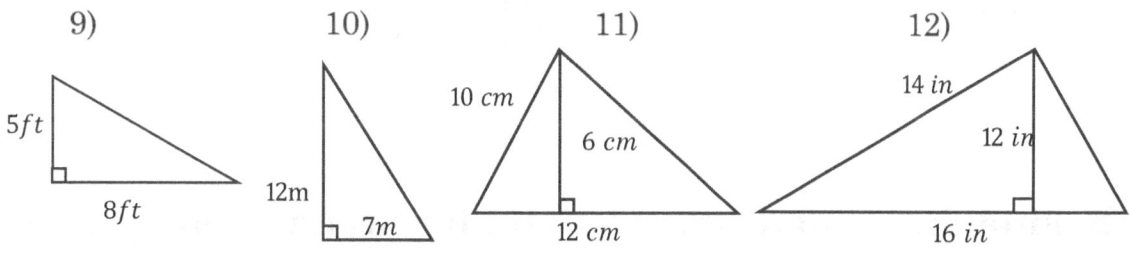

9)

5ft

8ft

10)

12m

7m

11)

10 cm   6 cm

12 cm

12)

14 in   12 in

16 in

✎ Find the perimeter or circumference of each shape.

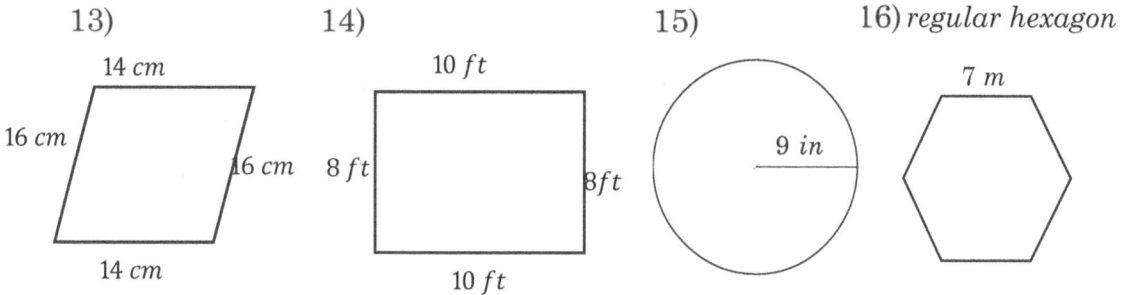

13)

14 cm

16 cm   16 cm

14 cm

14)

10 ft

8 ft   8ft

10 ft

15)

9 in

16) regular hexagon

7 m

✍ Find the area of each trapezoid.

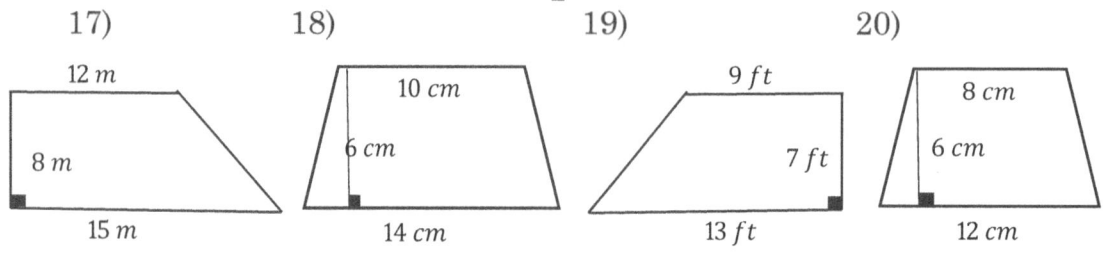

17)

12 m

8 m

15 m

18)

10 cm

6 cm

14 cm

19)

9 ft

7 ft

13 ft

20)

8 cm

6 cm

12 cm

✍ Find the volume of each cube.

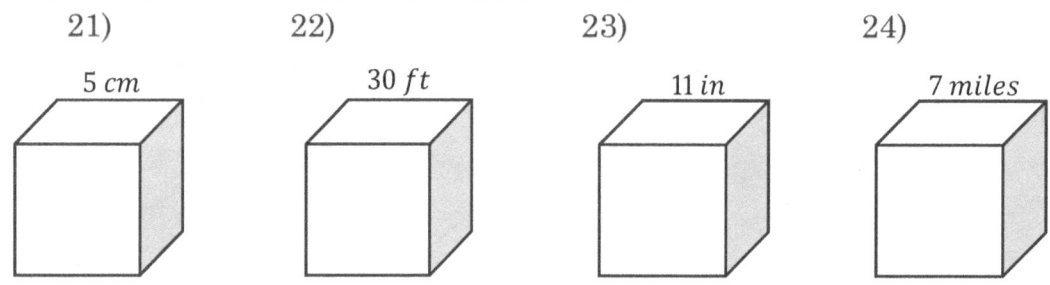

21)

5 cm

22)

30 ft

23)

11 in

24)

7 miles

✍ Find the volume of each Rectangular Prism.

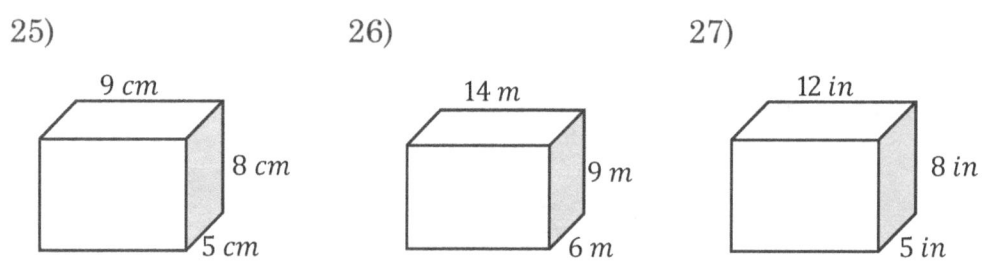

25)

9 cm

8 cm

5 cm

26)

14 m

9 m

6 m

27)

12 in

8 in

5 in

✍ Find the volume of each Cylinder. Round your answer to the nearest tenth. ($\pi = 3.14$)

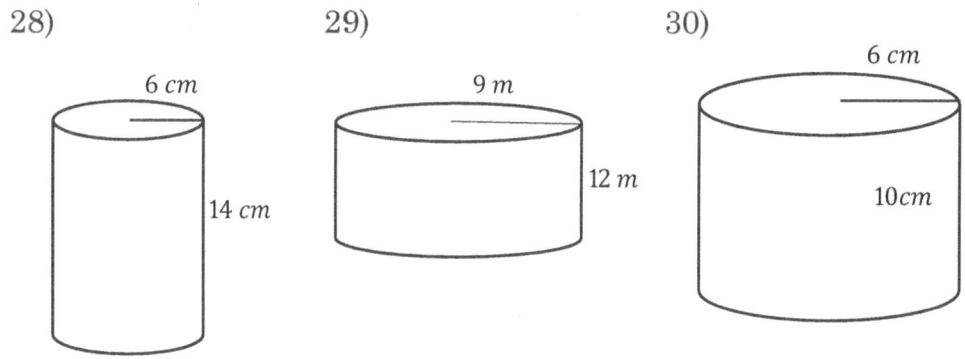

28)

6 cm

14 cm

29)

9 m

12 m

30)

6 cm

10 cm

✍ Find the values of the Given Data.

54) 9, 10, 9, 8, 11   55) 6, 8, 1, 4, 7, 6, 10

   Mode: _____   Range: _____   Mean: _____   Median: _____

✍ The circle graph below shows all Bob's expenses for last month. Bob spent \$675 on his Rent last month.

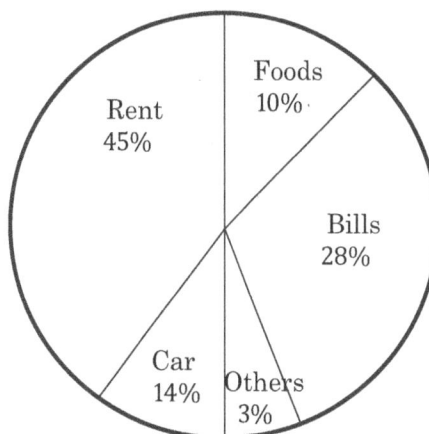

Bob's last month expenses

56) How much did Bob's total expenses last month? _____

57) How much did Bob spend for his bills last month? _____

58) How much did Bob spend for his car last month? _____

✍ Solve.

59) Bag A contains 9 red marbles and 6 green marbles. Bag B contains 6 black marbles and 9 orange marbles. What is the probability of selecting a red marble at random from bag A? What is the probability of selecting a black marble at random from Bag B?

_____   _____

60) Jason is planning for his vacation. He wants to go to museum, go to the beach, and play volleyball. How many different ways of ordering are there for him? _____

61) In how many ways can a team of 8 basketball players choose a captain and co-captain? _____

62) How many ways can you give 9 balls to your 12 friends? _____

Day 7: Answers

1) Use Pythagorean Theorem: $a^2 + b^2 = c^2$, $a = 6$ and $c = 10$, Then: $6^2 + b^2 = 10^2 \rightarrow 36 + b^2 = 100 \rightarrow b^2 = 100 - 36 = 64 \rightarrow b = \sqrt{64} = 8$

2) Use Pythagorean Theorem: $a^2 + b^2 = c^2$, $a = 15$ and $b = 8$, Then: $15^2 + 8^2 = c^2 \rightarrow 225 + 64 = c^2 \rightarrow c^2 = 289 \rightarrow c = \sqrt{289} = 17$

3) Use Pythagorean Theorem: $a^2 + b^2 = c^2$, $a = 12$ and $c = 13$, Then: $12^2 + b^2 = 13^2 \rightarrow 144 + b^2 = 169 \rightarrow b^2 = 169 - 144 = 25 \rightarrow b = \sqrt{25} = 5$

4) Use Pythagorean Theorem: $a^2 + b^2 = c^2$, $a = 3$ and $b = 4$, Then: $3^2 + 4^2 = c^2 \rightarrow 9 + 16 = c^2 \rightarrow c^2 = 25 \rightarrow c = \sqrt{25} = 5$

5) In any triangle, the sum of all angles is 180 degrees. Then: $85 + 40 + x = 180 \rightarrow 125 + x = 180 \rightarrow x = 180 - 125 = 55$

6) $58 + 42 + x = 180 \rightarrow 100 + x = 180 \rightarrow x = 180 - 100 = 80$

7) $71 + 18 + x = 180 \rightarrow 89 + x = 180 \rightarrow x = 180 - 89 = 91$

8) $28 + 68 + x = 180 \rightarrow 96 + x = 180 \rightarrow x = 180 - 96 = 84$

9) Area $= \frac{1}{2}$(base$\times$height), base$= 8$ and height $= 5$; Area $= \frac{1}{2}(8 \times 5) = \frac{40}{2} = 20 \, ft^2$

10) Base $= 7 \, m$ and height $= 12 \, m$; Area $= \frac{1}{2}(12 \times 7) = \frac{84}{2} = 42 \, m^2$

11) Base $= 12 \, cm$ and height $= 6 \, cm$; Area $= \frac{1}{2}(12 \times 6) = \frac{72}{2} = 36 \, cm^2$

12) Base $= 16 \, in$ and height $= 12 \, in$; Area $= \frac{1}{2}(16 \times 12) = \frac{192}{2} = 96 \, in^2$

13) The perimeter of a parallelogram $= 2(l + w) \rightarrow l = 14 \, cm, w = 16 \, cm$. Then: $2(l + w) = 2(14 + 16) = 60 \, cm$

14) The perimeter of a rectangle $= 2(w + l) \rightarrow l = 10 \, ft, w = 8 \, ft$. Then: $2(l + w) = 2(10 + 8) = 36 \, ft$

15) Circumference of a circle $= 2\pi r \rightarrow r = 9 \, in$. Then: $2\pi \times 9 = 18 \times 3.14 = 56.52 \, in$

16) The perimeter of a regular hexagon $= 6a \rightarrow a = 7 \, m$. Then: $6 \times 7 = 42 \, m$

17) The area of a trapezoid $= \frac{1}{2}h(b_1 + b_2) \rightarrow b_1 = 12\,m$, $b_2 = 15\,m$ and $h = 8m$.
Then: $A = \frac{1}{2}(8)(12 + 15) = 108\,m^2$

18) $b_1 = 10\,cm$, $b_2 = 14\,cm$ and $h = 6\,cm$. Then: $A = \frac{1}{2}(6)(10 + 14) = 72\,cm^2$

19) $b_1 = 9\,ft$, $b_2 = 13\,ft$ and $h = 7\,ft$. Then: $A = \frac{1}{2}(7)(9 + 13) = 77\,ft^2$

20) $b_1 = 8\,cm$, $b_2 = 12\,cm$ and $h = 6\,cm$. Then: $A = \frac{1}{2}(6)(8 + 12) = 60\,cm^2$

21) The volume of a cube $= (one\ side)^3 = (5)^3 = 125\,cm^3$

22) The volume of a cube $= (one\ side)^3 = (30)^3 = 27{,}000\,ft^3$

23) The volume of a cube $= (one\ side)^3 = (11)^3 = 1{,}331\,in^3$

24) The volume of a cube $= (one\ side)^3 = (7)^3 = 343\,miles^3$

25) The volume of a Rectangular prism $= l \times w \times h \rightarrow l = 9\,cm, w = 5\,cm, h = 8\,cm$.
Then: $V = 9 \times 5 \times 8 = 360\,cm^3$

26) $V = l \times w \times h \rightarrow l = 14\,m, w = 6\,m, h = 9\,m$. Then: $V = 14 \times 6 \times 9 = 756\,m^3$

27) $V = l \times w \times h \rightarrow l = 12\,in, w = 5\,in, h = 8\,in$. Then: $V = 12 \times 5 \times 8 = 480\,in^3$

28) Volume of a Cylinder $= \pi(r)^2 \times h \rightarrow r = 6\,cm, h = 14\,cm$
Then: $\pi(6)^2 \times 14 = 3.14 \times 36 \times 14 = 1{,}582.56 \approx 1{,}582.6$

29) Volume of a Cylinder $= \pi(r)^2 \times h \rightarrow r = 9\,m, h = 12\,m$
Then: $\pi(9)^2 \times 12 = 3.14 \times 81 \times 12 = 3{,}052.08 \approx 3{,}052.1$

30) Volume of a Cylinder $= \pi(r)^2 \times h \rightarrow r = 6\,cm, h = 10\,cm$
Then: $\pi(6)^2 \times 10 = 3.14 \times 36 \times 10 = 1{,}130$

31) Mode = the value in the list that appears most often = 9

Range: the difference of the largest value and smallest value. Then:
$$11 - 8 = 3$$

32) Mean: $\dfrac{sum\ of\ the\ data}{total\ number\ of\ data\ entires} = \dfrac{6+8+1+4+7+6+10}{7} = \dfrac{42}{7} = 6$

Median: is the middle number of a group of numbers arranged in order by size. Write the numbers in order: 1, 4, 6, 7, 8, 10. Then: Median $= \dfrac{6+7}{2} = 6.5$

33) Bob's rent = $675

Total expenses $\rightarrow$ 45% of total expenses = rent $\rightarrow$

Total expenses = rent $\div$ 0.45 = \$675 $\div$ 0.45 = \$1,500

34) Bills = 28% $\times$ total expenses = 0.28 $\times$ \$1,500 = \$420

35) Car = 14% $\times$ total expenses = 0.14 $\times$ \$1,500 = 210

36) Probability of pulling out a red marble from bag

$$A = \frac{numbers\ of\ red\ marbles}{total\ of\ marbles} = \frac{9}{6+9} = \frac{9}{15} = 0.6 = 60\%$$

Probability of pulling out a black marble from bag

$$B = \frac{numbers\ of\ black\ marbles}{total\ of\ marbles} = \frac{6}{6+9} = \frac{6}{15} = 0.4 = 40\%$$

37) Jason has 3 choices. Therefore, number of different ways of ordering the events is: $3 \times 2 \times 1 = 6$

38) This is a permutation problem. (order is important) Then: $nP_k = \frac{n!}{(n-k)!} \to n = 8, k = 2 \to \frac{8!}{(8-2)!} = \frac{8!}{6!} = \frac{8 \times 7 \times 6!}{6!} = 56$

39) This is a combination problem (order doesn't matter) $nC_r = \frac{n!}{r!\,(n-r)!} \to$

$$n = 12, r = 9 \to nC_r = \frac{12!}{9!\,(12-9)!} = \frac{12!}{9!\,3!} = \frac{12 \times 11 \times 10 \times 9!}{3 \times 2 \times 1 \times 9!} = \frac{1,320}{6} = 220$$

DAY 8 Functions and Quadratic

Math topics that you'll learn in this chapter:

1. Function Notation and Evaluation
2. Adding and Subtracting Functions
3. Multiplying and Dividing Functions
4. Compositions of Functions
5. Solving a Quadratic Equation
6. Graphing Quadratic Functions
7. Solving Quadratic Inequalities
8. Graphing Quadratic Inequalities

99

Function Notation and Evaluation

☆ Functions are mathematical operations that assign unique outputs to given inputs.

☆ Function notation is the way a function is written. It is meant to be a precise way of giving information about the function without a rather lengthy written explanation.

☆ The most popular function notation is $f(x)$ which is read "f of x". Any letter can name a function. for example: $g(x)$, $h(x)$, etc.

☆ To evaluate a function, plug in the input (the given value or expression) for the function's variable (place holder, x).

Examples:

Example 1. Evaluate: $f(x) = 2x + 9$, find $f(5)$

Solution: Substitute x with 5:
Then: $f(x) = 2x + 9 \rightarrow f(5) = 2(5) + 9 \rightarrow f(5) = 19$

Example 2. Evaluate: $w(x) = 5x - 4$, find $w(2)$.

Solution: Substitute x with 2:
Then: $w(x) = 5x - 4 \rightarrow w(2) = 5(2) - 4 = 10 - 4 = 6$

Example 3. Evaluate: $f(x) = 5x^2 + 8$, find $f(-2)$.

Solution: Substitute x with -2:
Then: $f(x) = 5x^2 + 8 \rightarrow f(-2) = 5(-2)^2 + 8 \rightarrow f(-2) = 20 + 8 = 28$

Example 4. Evaluate: $h(x) = 3x^2 - 4$, find $h(3a)$.

Solution: Substitute x with $3a$:
Then: $h(x) = 3x^2 - 4 \rightarrow h(3a) = 3(3a)^2 - 4 \rightarrow h(3a) = 3(9a^2) - 4 = 27a^2 - 4$

Adding and Subtracting Functions

☆ Just like we can add and subtract numbers and expressions, we can add or subtract functions and simplify or evaluate them. The result is a new function.

☆ For two functions $f(x)$ and $g(x)$, we can create two new functions:

$$(f + g)(x) = f(x) + g(x) \text{ and } (f - g)(x) = f(x) - g(x)$$

Examples:

Example 1. $g(x) = 4x - 3$, $f(x) = x + 6$, Find: $(g + f)(x)$

Solution: $(g + f)(x) = g(x) + f(x)$
Then: $(g + f)(x) = (4x - 3) + (x + 6) = 4x - 3 + x + 6 = 5x + 3$

Example 2. $f(x) = 2x - 7$, $g(x) = x - 9$, Find: $(f - g)(x)$

Solution: $(f - g)(x) = f(x) - g(x)$
Then: $(f - g)(x) = (2x - 7) - (x - 9) = 2x - 7 - x + 9 = x + 2$

Example 3. $g(x) = 2x^2 + 6$, $f(x) = x - 3$, Find: $(g + f)(x)$

Solution: $(g + f)(x) = g(x) + f(x)$
Then: $(g + f)(x) = (2x^2 + 6) + (x - 3) = 2x^2 + x + 3$

Example 4. $f(x) = 2x^2 - 1$, $g(x) = 4x + 3$, Find: $(f - g)(2)$

Solution: $(f - g)(x) = f(x) - g(x)$
Then: $(f - g)(x) = (2x^2 - 1) - (4x + 3) = 2x^2 - 1 - 4x - 3 = 2x^2 - 4x - 4$
Substitute x with 2: $(f - g)(2) = 2(2)^2 - 4(2) - 4 = 8 - 8 - 4 = -4$

Multiplying and Dividing Functions

☆ Just like we can multiply and divide numbers and expressions, we can multiply and divide two functions and simplify or evaluate them.

☆ For two functions $f(x)$ and $g(x)$, we can create two new functions:

$$(f.g)(x) = f(x).g(x) \text{ and } \left(\frac{f}{g}\right)(x) = \frac{f(x)}{g(x)}$$

Examples:

Example 1. $g(x) = x + 2$, $f(x) = x + 5$, Find: $(g.f)(x)$

Solution:

$(g.f)(x) = g(x).f(x) = (x + 2)(x + 5) = x^2 + 5x + 2x + 10 = x^2 + 7x + 10$

Example 2. $f(x) = x + 4$, $h(x) = x - 16$, Find: $\left(\frac{f}{h}\right)(x)$

Solution: $\left(\frac{f}{h}\right)(x) = \frac{f(x)}{h(x)} = \frac{x+4}{x-16}$

Example 3. $g(x) = x + 5$, $f(x) = x - 2$, Find: $(g.f)(3)$

Solution: $(g.f)(x) = g(x).f(x) = (x + 5)(x - 2) = x^2 - 2x + 5x - 10$

$$g(x).f(x) = x^2 + 3x - 10$$

Substitute x with 3: $(g.f)(3) = (3)^2 + 3(3) - 10 = 9 + 9 - 10 = 8$

Example 4. $f(x) = 2x + 2$, $h(x) = x - 3$, Find: $\left(\frac{f}{h}\right)(4)$

Solution: $\left(\frac{f}{h}\right)(x) = \frac{f(x)}{h(x)} = \frac{2x+2}{x-3}$

Substitute x with 4: $\left(\frac{f}{h}\right)(4) = \frac{2x+2}{x-3} = \frac{2(4)+2}{4-3} = \frac{10}{1} = 10$

Composition of Functions

☆ "Composition of functions" simply means combining two or more functions in a way where the output from one function becomes the input for the next function.

☆ The notation used for composition is: $(fog)(x) = f(g(x))$ and is read

"f composed with g of x" or "f of g of x".

Examples:

Example 1. Using $f(x) = x + 5$ and $g(x) = 7x$, find: $(fog)(x)$

Solution: $(fog)(x) = f(g(x))$. Then: $(fog)(x) = f(g(x)) = f(7x)$

Now find $f(7x)$ by substituting x with $7x$ in $f(x)$ function.

Then: $f(x) = x + 5$; $(x \rightarrow 7x) \rightarrow f(7x) = 7x + 5$

Example 2. Using $f(x) = 2x - 3$ and $g(x) = x - 5$, find: $(gof)(3)$

Solution: $(fog)(x) = f(g(x))$. Then: $(gof)(x) = g(f(x)) = g(2x - 3)$,

Now substitute x in $g(x)$ by $(2x - 3)$.

Then: $g(2x - 3) = (2x - 3) - 5 = 2x - 8$

Substitute x with 3: $(gof)(3) = g(f(x)) = 2x - 8 = 2(3) - 8 = -2$

Example 3. Using $f(x) = 3x^2 - 7$ and $g(x) = 2x + 1$, find: $f(g(2))$

Solution: First, find $g(2)$: $g(x) = 2x + 1 \rightarrow g(2) = 2(2) + 1 = 5$

Then: $f(g(2)) = f(5)$. Now, find $f(5)$ by substituting x with 5 in $f(x)$ function.

$f(g(2)) = f(5) = 3(5)^2 - 7 = 3(25) - 7 = 68$

Solving a Quadratic Equation

☆ Write the equation in the form of: $ax^2 + bx + c = 0$

☆ Factorize the quadratic, set each factor equal to zero and solve.

☆ Use quadratic formula if you couldn't factorize the quadratic.

☆ Quadratic formula: $x = \frac{-b \pm \sqrt{b^2 - 4ac}}{2a}$

Examples:

Find the solutions of each quadratic function.

Example 1. $x^2 + 7x + 12 = 0$

Solution: Factor the quadratic by grouping. We need to find two numbers whose sum is 7 (from $7x$) and whose product is 12. Those numbers are 3 and 4. Then: $x^2 + 7x + 12 = 0 \rightarrow x^2 + 3x + 4x + 12 = 0 \rightarrow (x^2 + 3x) + (4x + 12) = 0$, Now, find common factors: $(x^2 + 3x) = x(x + 3)$ and $(4x + 12) = 4(x + 3)$. We have two expressions $(x^2 + 3x)$ and $(4x + 12)$ and their common factor is $(x + 3)$. Then: $(x^2 + 3x) + (4x + 12) = 0 \rightarrow x(x + 3) + 4(x + 3) = 0 \rightarrow (x + 3)(x + 4) = 0$.

The product of two expressions is 0. Then:

$(x + 3) = 0 \rightarrow x = -3$ or $(x + 4) = 0 \rightarrow x = -4$

Example 2. $x^2 + 5x + 6 = 0$

Solution: Use quadratic formula: $x_{1,2} = \frac{-b \pm \sqrt{b^2 - 4ac}}{2a}$, $a = 1, b = 5$ and $c = 6$

Then: $x = \frac{-5 \pm \sqrt{5^2 - 4 \times 1(6)}}{2(1)}$, $x_1 = \frac{-5 + \sqrt{5^2 - 4 \times 1(6)}}{2(1)} = -2$, $x_2 = \frac{-5 - \sqrt{5^2 - 4 \times 1(6)}}{2(1)} = -3$

Graphing Quadratic Functions

☆ Quadratic functions in vertex form: $y = a(x - h)^2 + k$ where (h, k) is the vertex of the function. The axis of symmetry is $x = h$

☆ Quadratic functions in standard form: $y = ax^2 + bx + c$ where $x = -\frac{b}{2a}$ is the value of x in the vertex of the function.

☆ To graph a quadratic function, first find the vertex, then substitute some values for x and solve for y. (Remember that the graph of a quadratic function is a U-shaped curve and it is called "parabola".)

Example:

Example 1. Sketch the graph of $y = (x + 2)^2 - 3$

Solution: Quadratic functions in vertex form: $y = a(x - h)^2 + k$ and (h, k) is the vertex. Then, the vertex of $y = (x + 2)^2 - 3$ is $(-2, -3)$.
Substitute zero for x and solve for y:
$y = (0 + 2)^2 - 3 = 1$.
The y Intercept is $(0, 1)$.
Now, you can simply graph the quadratic function. Notice that quadratic function is a U-shaped curve.

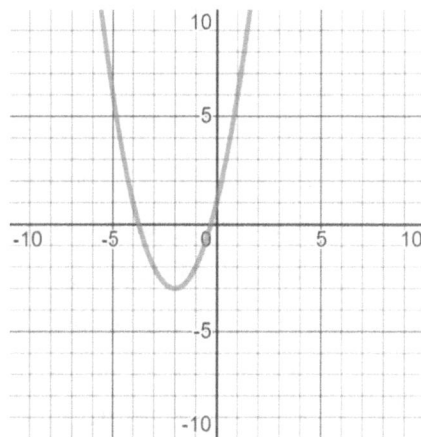

Example 2. Sketch the graph of $y = (x + 3)^2 - 4$

Solution: Quadratic functions in vertex form: $y = a(x - h)^2 + k$ and (h, k) is the vertex. Then, the vertex of $y = (x + 3)^2 - 4$ is $(-3, -4)$.
Substitute zero for x and solve for y: $y = (0 + 3)^2 - 4 = 5$.
The y Intercept is $(0, 5)$.
Now, you can simply graph the quadratic function. Notice that quadratic function is a U-shaped curve.

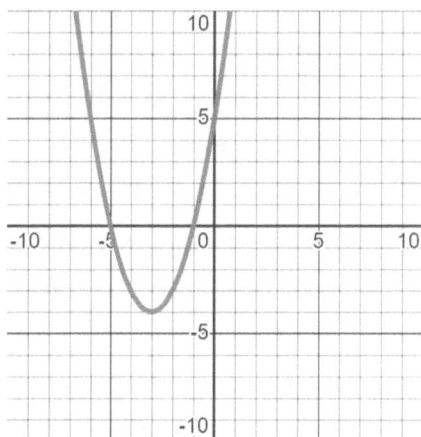

Solving Quadratic Inequalities

☆ A quadratic inequality is one that can be written in the standard form of

$ax^2 + bx + c > 0$ (or substitute $<, \leq,$ or $\geq$ for $>$).

☆ Solving a quadratic inequality is like solving equations. We need to find the solutions (the zeroes).

☆ To solve quadratic inequalities, first find quadratic equations. Then choose a test value between zeroes. Finally, find interval(s), such as > 0 or < 0.

Examples:

Example 1. Solve quadratic inequality. $x^2 + x - 6 > 0$

Solution: First solve $x^2 + x - 6 = 0$ by factoring. Then: $x^2 + x - 6 = 0 \rightarrow$ $(x - 2)(x + 3) = 0$. The product of two expressions is 0. Then: $(x - 2) = 0 \rightarrow x = 2$ or $(x + 3) = 0 \rightarrow x = -3$. Now, choose a value between 2 and -3. Let's choose 0. Then: $x = 0 \rightarrow x^2 + x - 6 > 0 \rightarrow (0)^2 + (0) - 6 > 0 \rightarrow -6 > 0$

-6 is not greater than 0. Therefore, all values between 2 and -3 are NOT the solution of this quadratic inequality. The solution is: $x > 2$ and $x < -3$. To represent the solution, we can use interval notation, in which solution sets are indicated with parentheses or brackets. The solutions $x > 2$ and

$x < -3$ represented as: $(-\infty, -3) \cup (2, \infty)$

Example 2. Solve quadratic inequality. $x^2 + 2x - 15 > 0$

Solution: First solve $x^2 + 2x - 15 = 0$ by factoring. Then: $x^2 + 2x - 15 = 0 \rightarrow$ $(x - 3)(x + 5) = 0$. The product of two expressions is 0. Then: $(x - 3) = 0 \rightarrow x = $ 3 or $(x + 5) = 0 \rightarrow x = -5$. Now, choose a value between 3 and -5. Let's choose 0. Then: $x = 0 \rightarrow x^2 + 2x - 15 > 0 \rightarrow (0)^2 + 2(0) - 15 > 0 \rightarrow -15 > 0$

-15 is not greater than 0. Therefore, all values between 3 and -5 are NOT the solution of this quadratic inequality. The solution is: $x > 3$ and $x < -5$. To represent the solution, we can use interval notation, in which solution sets are indicated with parentheses or brackets. The solutions $x > 3$ and

$x < -5$ represented as: $(-\infty, -5) \cup (3, \infty)$

Graphing Quadratic Inequalities

☆ A quadratic inequality is in the form

$y > ax^2 + bx + c$ (or substitute $<$, $\leq$, or $\geq$ for $>$).

☆ To graph a quadratic inequality, start by graphing the quadratic parabola. Then fill in the region either inside or outside of it, depending on the inequality.

☆ Choose a testing point and check the solution section.

Example:

Example 1. Sketch the graph of $y > 2x^2$

Solution: First, graph the quadratic $y = 2x^2$
Since the inequality sing is $>$, we need to use dash lines. Now, choose a testing point inside the parabola. Let's choose $(0, 2)$.
$y > 2x^2 \rightarrow 2 > 2(0)^2 \rightarrow 2 > 0$

This is true. So, inside the parabola is the solution section.

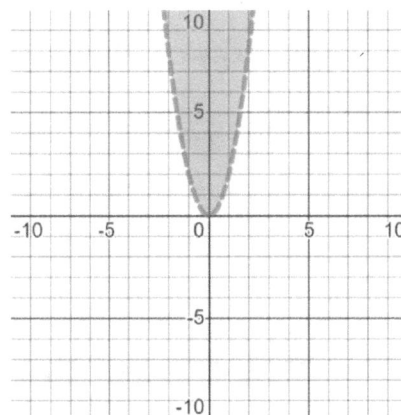

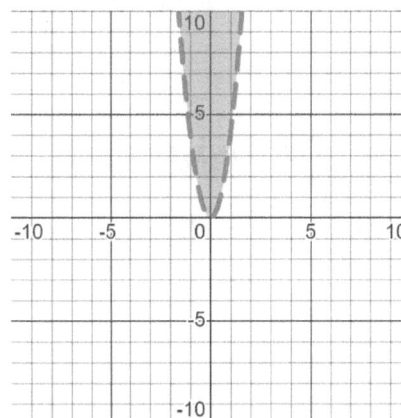

Example 2. Sketch the graph of $y > 4x^2$

Solution: First, graph the quadratic $y = 4x^2$
Since the inequality sing is $>$, we need to use dash lines.

Now, choose a testing point inside the parabola. Let's choose $(0, 4)$.

$y > 4x^2 \rightarrow 4 > 4(0)^2 \rightarrow 4 > 0$

This is true. So, inside the parabola is the solution section.

Day 8: Practices

✍ Evaluate each function.

1) $g(n) = 3n + 7$, find $g(3)$

_____

2) $h(x) = 4n - 7$, find $h(5)$

_____

3) $y(n) = 12 - 3n$, find $y(8)$

_____

4) $b(n) = -10 - 5n$, find $b(8)$

_____

5) $g(x) = -7x + 6$, find $g(-3)$

_____

6) $k(n) = -4n + 5$, find $k(-5)$

_____

7) $w(n) = -3n - 6$, find $w(-4)$

_____

8) $z(n) = 14 - 2n$, find $n(3)$

_____

✍ Perform the indicated operation.

9) $f(x) = 2x + 3$

$g(x) = x + 4$

Find $(f - g)(x)$

_____

10) $g(a) = -3a + 2$

$h(a) = a^2 - 4$

Find $(h + g)(3)$

_____

✍ Perform the indicated operation.

11) $g(x) = x - 4$

$f(x) = x + 3$

Find $(g.f)(2)$

_____

12) $f(x) = x + 2$

$h(x) = x - 5$

Find $\left(\frac{f}{h}\right)(-3)$

_____

✍ Using $f(x) = 2x + 5$ and $g(x) = 2x - 3$, find:

13) $g(f(2)) =$ ____

14) $g(f(-2)) =$ ____

15) $f(f(1)) =$ ____

16) $f(f(-1)) =$ ____

17) $f(g(4)) =$ ____

18) $f(g(-3)) =$ ____

✍ **Solve each equation by factoring or using the quadratic formula.**

19) $x^2 - 4x - 32 = 0$

21) $x^2 + 17x + 72 = 0$

20) $x^2 - 2x - 63 = 0$

22) $x^2 + 14x + 48 = 0$

✍ **Sketch the graph of each function.**

23) $y = (x + 1)^2 - 2$

24) $y = (x - 1)^2 + 3$

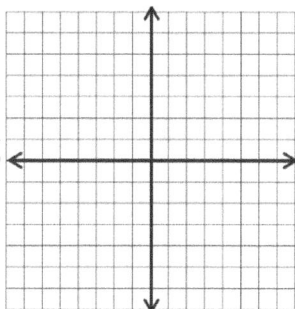

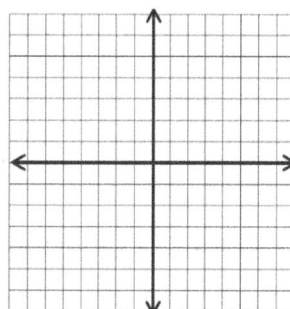

✍ **Solve each quadratic inequality.**

25) $x^2 - 4 < 0$

27) $x^2 - 5x - 6 < 0$

26) $x^2 - 9 > 0$

28) $x^2 + 8x - 20 > 0$

✍ **Sketch the graph of each quadratic inequality.**

29) $y < -2x^2$

30) $y > 3x^2$

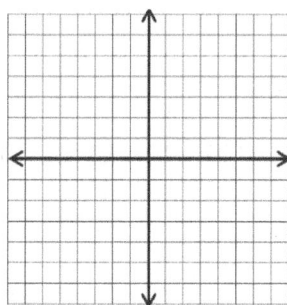

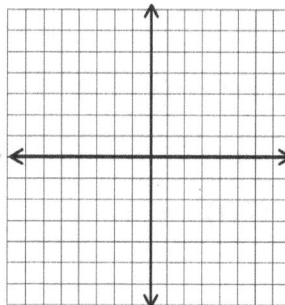

Day 8: Answers

1) $g(n) = 3n + 7 \rightarrow g(3) = 3(3) + 7 = 9 + 7 = 16$

2) $h(x) = 4n - 7 \rightarrow h(5) = 4(5) - 7 = 20 - 7 = 13$

3) $y(n) = 12 - 3n \rightarrow y(8) = 12 - 3(8) = 12 - 24 = -12$

4) $b(n) = -10 - 5n \rightarrow b(8) = -10 - 5(8) = -10 - 40 = -50$

5) $g(x) = -7x + 6 \rightarrow g(-3) = -7(-3) + 6 = 21 + 6 = 27$

6) $k(n) = -4n + 5 \rightarrow k(-5) = -4(-5) + 5 = 20 + 5 = 25$

7) $w(n) = -3n - 6 \rightarrow w(-4) = -3(-4) - 6 = 12 - 6 = 6$

8) $z(n) = 14 - 2n \rightarrow n(3) = 14 - 2(3) = 14 - 6 = 8$

9) $(f - g)(x) = f(x) - g(x) = 2x + 3 - (x + 4) = 2x + 3 - x - 4 = x - 1$

10) $(h + g)(3) = h(a) + g(a) = a^2 - 4 - 3a + 2 = a^2 - 3a - 2 = (3)^2 - 3(3) - 2 = 9 - 9 - 2 = -2$

11) $(g.f)(2) = f(x) \times g(x) = (x - 4) \times (x + 3)$

$= (x \times x) + (x \times 3) + (-4 \times x) + (-4 \times 3) = x^2 + 3x - 4x - 12 = x^2 - x - 12 = 2^2 - 2 - 12 = 4 - 2 - 12 = -10$

12) $\left(\frac{f}{h}\right)(-3) = \frac{f(x)}{h(x)} = \frac{x+2}{x-5} = \frac{(-3)+2}{(-3)-5} = \frac{-3+2}{-3-5} = \frac{-1}{-8} = \frac{1}{8}$

13) First, find $f(2)$: $f(x) = 2x + 5 \rightarrow f(2) = 2(2) + 5 = 9$

Then: $g(f(2)) = g(9)$. Now, find $g(9)$ by substituting x with 9 in $g(x)$ function. $g(f(2)) = g(9) = 2(9) - 3 = 18 - 3 = 15$

14) First, find $f(-2)$: $f(x) = 2x + 5 \rightarrow f(-2) = 2(-2) + 5 = -4 + 5 = 1$

Then: $g(f(-2)) = g(1)$. Now, find $g(1)$ by substituting x with 1 in $g(x)$ function. $g(f(-2)) = g(1) = 2(1) - 3 = 2 - 3 = -1$

15) First, find $f(1)$: $f(x) = 2x + 5 \rightarrow f(1) = 2(1) + 5 = 2 + 5 = 7$

Then: $f(f(1)) = f(7)$. Now, find $f(7)$ by substituting x with 7 in $f(x)$ function. $f(f(1)) = f(7) = 2(7) + 5 = 14 + 5 = 19$

16) First, find $f(-1)$: $f(x) = 2x + 5 \rightarrow f(-1) = 2(-1) + 5 = -2 + 5 = 3$

Then: $f(f(-1)) = f(3)$. Now, find $f(3)$ by substituting x with 3 in $f(x)$

function. $f(g(-1)) = f(3) = 2(3) + 5 = 6 + 5 = 11$

17) First, find $g(4)$: $g(x) = 2x - 3 \rightarrow g(4) = 2(4) - 3 = 8 - 3 = 5$

Then: $f(g(4)) = f(5)$. Now, find $f(5)$ by substituting x with 5 in $f(x)$

function. $f(g(4)) = f(5) = 2(5) + 5 = 10 + 5 = 15$

18) First, find $g(-3)$: $g(x) = 2x - 3 \rightarrow g(-3) = 2(-3) - 3 = -6 - 3 = -9$

Then: $f(g(-3)) = f(-9)$. Now, find $f(-9)$ by substituting x with -9 in

$f(x)$. function. $f(g(-3)) = f(-9) = 2(-9) + 5 = -18 + 5 = -13$

19) $8, -4$ (Factor the quadratic by grouping. We need to find two numbers whose sum is -4 (from $-4x$) and whose product is -32. Those numbers are -8 and 4. Then: $x^2 - 4x - 32 = 0 \rightarrow x^2 - 8x + 4x - 32 = 0 \rightarrow (x^2 - 8x) + (4x - 32) = 0$, Now, find common factors: $(x^2 - 8x) = x(x - 8)$ and $(4x - 32) = 4(x - 8)$. We have two expressions $(x^2 - 8x)$ and $(4x - 32)$ and their common factor is $(x - 8)$. Then: $(x^2 - 8x) + (4x - 32) = 0 \rightarrow x(x - 8) + 4(x - 8) = 0 \rightarrow (x - 8)(x + 4) = 0$. The product of two expressions is 0. Then: $(x - 8) = 0 \rightarrow x = 8$ or $(x + 4) = 0 \rightarrow x = -4$)

20) $9, -7$ (Factor the quadratic by grouping. We need to find two numbers whose sum is -2 (from $-2x$) and whose product is -63. Those numbers are -9 and 7. Then: $x^2 - 2x - 63 = 0 \rightarrow x^2 - 9x + 7x - 63 = 0 \rightarrow (x^2 - 9x) + (7x - 63) = 0$, Now, find common factors: $(x^2 - 9x) = x(x - 9)$ and $(7x - 63) = 7(x - 9)$. We have two expressions $(x^2 - 9x)$ and $(7x - 63)$ and their common factor is $(x - 9)$. Then: $(x^2 - 9x) + (7x - 63) = 0 \rightarrow x(x - 9) + 7(x - 9) = 0 \rightarrow (x - 9)(x + 7) = 0$. The product of two expressions is 0. Then: $(x - 9) = 0 \rightarrow x = 9$ or $(x + 7) = 0 \rightarrow x = -7$)

21) $-9, -8$ (Factor the quadratic by grouping. We need to find two numbers whose sum is 17 (from $17x$) and whose product is 72. Those numbers are 9 and 8 . Then: $x^2 + 17x + 72 = 0 \rightarrow x^2 + 9x + 8x + 72 = 0 \rightarrow (x^2 + 9x) + (8x + 72) = 0$, Now, find common factors: $(x^2 + 9x) = x(x + 9)$ and $(8x + 72) = 8(x + 9)$. We have two expressions $(x^2 + 9x)$ and $(8x + 72)$ and their common factor is $(x + 9)$. Then: $(x^2 + 9x) + (8x + 72) = 0 \rightarrow x(x + 9) + 8(x + 9) = 0 \rightarrow (x + 9)(x + 8) = 0$. The product of two expressions is 0. Then: $(x + 9) = 0 \rightarrow x = -9$ or $(x + 8) = 0 \rightarrow x = -8$)

22) $-8, -6$ (Factor the quadratic by grouping. We need to find two numbers whose sum is 14 (from $14x$) and whose product is 48. Those numbers are 6 and 8 . Then: $x^2 + 14x + 48 = 0 \rightarrow x^2 + 8x + 6x + 48 = 0 \rightarrow (x^2 + 8x) + (6x + 48) = 0$, Now, find common factors: $(x^2 + 8x) = x(x + 8)$ and $(6x + 48) = 6(x + 8)$. We have two expressions $(x^2 + 8x)$ and $(6x + 48)$ and their common factor is $(x + 8)$. Then: $(x^2 + 8x) + (6x + 48) = 0 \rightarrow x(x + 8) + 6(x + 8) = 0 \rightarrow (x + 8)(x + 6) = 0$. The product of two expressions is 0. Then: $(x + 8) = 0 \rightarrow x = -8$ or $(x + 6) = 0 \rightarrow x = -6$)

23) Quadratic functions in vertex form: $y = a(x - h)^2 + k$ and (h, k) is the vertex. Then, the vertex of $y = (x + 1)^2 - 2$ is $(-1, -2)$. Substitute zero for x and solve for y: $y = (0 + 1)^2 - 2 = -1$. The y Intercept is $(0, -1)$. Now, you can simply graph the quadratic function. Notice that quadratic function is a U-shaped curve.

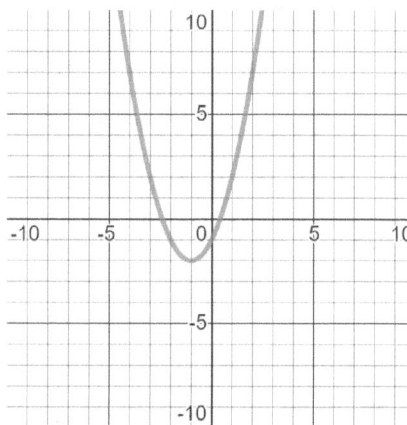

24) Quadratic functions in vertex form: $y = a(x - h)^2 + k$ and (h, k) is the vertex. Then, the vertex of $y = (x - 1)^2 + 3$ is $(1, 3)$. Substitute zero for x and solve for y: $y = (0 - 1)^2 + 3 = 4$. The y Intercept is $(0, 4)$. Now, you can simply graph the quadratic function. Notice that quadratic function is a U-shaped curve.

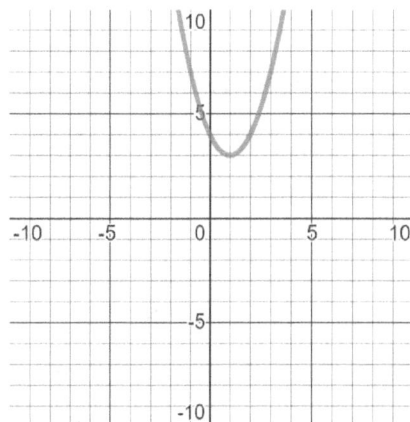

25) $-2 < x < 2$ (First solve: $x^2 - 4 = 0$, Factor: $x^2 - 4 = 0 \rightarrow (x - 2)(x + 2) = 0$. -2 and 2 are the solutions. Choose a point between -2 and 2. Let's choose 0. Then: $x = 0 \rightarrow x^2 - 4 < 0 \rightarrow (0)^2 - 4 < 0 \rightarrow -4 < 0$. This is true. So, the solution is: $-2 < x < 2$ (using interval notation the solution is: $(-2, 2)$)

26) $x < -3$ or $x > 3$ (First solve: $x^2 - 9 = 0$, Factor: $x^2 - 9 = 0 \rightarrow (x - 3)(x + 3) = 0$. -3 and 3 are the solutions. Choose a point between -3 and 3. Let's choose 0. Then: $x = 0 \rightarrow x^2 - 9 > 0 \rightarrow (0)^2 - 9 > 0 \rightarrow -9 > 0$. This is NOT true. So, the solution is: $x < -3$ or $x > 3$ (using interval notation the solution is: $(-\infty, -3) \cup (3, \infty)$)

27) $-1 < x < 6$ (First solve: $x^2 - 5x - 6 = 0$, Factor: $x^2 - 5x - 6 = 0 \rightarrow (x - 6)(x + 1) = 0$. -1 and 6 are the solutions. Choose a point between -1 and 6. Let's choose 0. Then: $x = 0 \rightarrow x^2 - 5x - 6 < 0 \rightarrow (0)^2 - 4(0) - 6 < 0 \rightarrow -6 < 0$. This is true. So, the solution is: $-1 < x < 6$ (using interval notation the solution is: $(-1, 6)$)

28) $x < -10$ or $x > 2$ (First solve: $x^2 + 8x - 20 = 0$, Factor: $x^2 + 8x - 20 = 0 \rightarrow (x + 10)(x - 2) = 0$. -10 and 2 are the solutions. Choose a point between -10 and 2. Let's choose 0. Then: $x = 0 \rightarrow x^2 + 8x - 20 > 0 \rightarrow (0)^2 + 8(0) - 20 > 0 \rightarrow -20 > 0$. This is NOT true. So, the solution is: $x < -10$ or $x > 2$ (using interval notation the solution is: $(-\infty, -10) \cup (2, \infty)$)

29) (First, graph the quadratic $y = -2x^2$. Since the inequality sing is $<$, we need to use dash lines. Now, choose a testing point inside the parabola. Let's choose $(0, -2)$. $y < -2x^2 \rightarrow -2 < -2(0)^2 \rightarrow -2 < 0$. This is true. So, inside the parabola is the solution section)

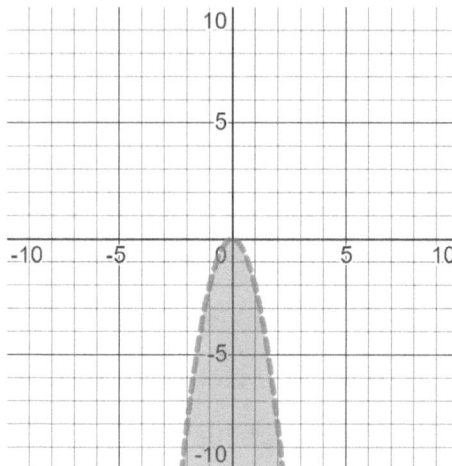

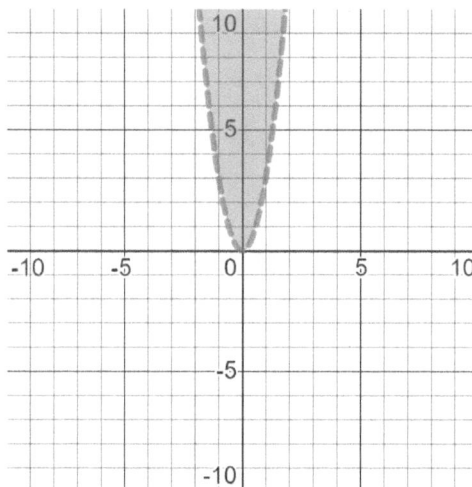

30) (First, graph the quadratic $y = 3x^2$. Since the inequality sing is $>$, we need to use dash lines. Now, choose a testing point inside the parabola. Let's choose $(0, 2)$. $y > 3x^2 \rightarrow 2 > 3(0)^2 \rightarrow 2 > 0$. This is true. So, inside the parabola is the solution section)

DAY 9

Complex Numbers and Radical Expressions

Math topics that you'll learn in this chapter:

115

Adding and Subtracting Complex Numbers

☆ A complex number is expressed in the form $a + bi$, where a and b are real numbers, and i, which is called an imaginary number, is a solution of the equation $x^2 = -1$

☆ For adding complex numbers:

$$(a + bi) + (c + di) = (a + c) + (b + d)i$$

☆ For subtracting complex numbers:

$$(a + bi) - (c + di) = (a - c) + (b - d)i$$

Examples:

Example 1. Solve: $(8 + 4i) + (6 - 2i)$

Solution: Remove parentheses: $(8 + 4i) + (6 - 2i) = 8 + 4i + 6 - 2i$
Combine like terms: $8 + 4i + 6 - 2i = 14 + 2i$

Example 2. Solve: $(-5 - 3i) - (2 + 4i)$

Solution: Remove parentheses by multiplying -1 to the second parentheses:
$(-5 - 3i) - (2 + 4i) = -5 - 3i - 2 - 4i$
Combine like terms: $-5 - 3i - 2 - 4i = -7 - 7i$

Example 3. Solve: $(12 + 6i) + (4 - 3i)$

Solution: Remove parentheses: $(12 + 6i) + (4 - 3i) = 12 + 6i + 4 - 3i$
Group like terms: $12 + 6i + 4 - 3i = 16 + 3i$

Multiplying and Dividing Complex Numbers

☆ You can use FOIL (First-Out-In-Last) method or the following rule to multiply imaginary numbers. Remember that: $i^2 = -1$

$$(a + bi) + (c + di) = (ac - bd) + (ad + bc)i$$

☆ To divide complex numbers, you need to find the conjugate of the denominator. Conjugate of $(a + bi)$ is $(a - bi)$.

☆ Dividing complex numbers: $\frac{a+bi}{c+di} = \frac{a+bi}{c+di} \times \frac{c-di}{c-di} = \frac{ac+bd}{c^2+d^2} + \frac{bc-ad}{c^2+d^2}i$

Examples:

Example 1. Solve: $\frac{6-2i}{2+i}$

Solution: The conjugate of $(2 + i)$ is $(2 - i)$. Use the rule for dividing complex numbers:

$\frac{a+bi}{c+di} = \frac{a+bi}{c+di} \times \frac{c-di}{c-di} = \frac{ac+bd}{c^2+d^2} + \frac{bc-ad}{c^2+d^2}i \rightarrow$

$\frac{6-2i}{2+i} \times \frac{2-i}{2-i} = \frac{6 \times (2) + (-2)(1)}{2^2 + (1)^2} + \frac{-2 \times 2 - (6)(1)}{2^2 + (1)^2}i = \frac{10}{5} + \frac{-10}{5}i = 2 - 2i$

Example 2. Solve: $(2 - 3i)(6 - 3i)$

Solution: Use the multiplication of imaginary numbers rule:
$(a + bi) + (c + di) = (ac - bd) + (ad + bc)i$
$\big(2 \times 6 - (-3)(-3)\big) + (2(-3) + (-3) \times 6)i = 3 - 24i$

Rationalizing Imaginary Denominators

☆ Step 1: Find the conjugate (it's the denominator with different sign between the two terms).

☆ Step 2: Multiply numerator and denominator by the conjugate.

☆ Step 3: Simplify if needed.

Examples:

Example 1. Solve: $\frac{4-3i}{6i}$

Solution: Multiply both numerator and denominator by $\frac{i}{i}$:

$$\frac{4-3i}{6i} = \frac{(4-3i)(i)}{6i(i)} = \frac{(4)(i)-(3i)(i)}{6(i^2)} = \frac{4i-3i^2}{6(-1)} = \frac{4i-3(-1)}{-6} = \frac{4i}{-6} + \frac{3}{-6} = -\frac{1}{2} - \frac{2}{3}i$$

Example 2. Solve: $\frac{6i}{2-i}$

Solution: Multiply both numerator and denominator by the conjugate

$$\frac{2+i}{2+i}: \frac{6i(2+i)}{(2-i)(2+i)} = \text{Apply complex arithmetic rule: } (a+bi)(a-bi) = a^2 + b^2$$

$$2^2 + (-1)^2 = 5, \text{ then: } \frac{6i(2+i)}{(2-i)(2+i)} = \frac{-6+12i}{5} = -\frac{6}{5} + \frac{12}{5}i$$

Example 3. Solve: $\frac{8-2i}{2i}$

Solution: Factor 2 from both sides: $\frac{8-2i}{2i} = \frac{2(4-i)}{2i}$, divide both sides by 2:

$$\frac{2(4-i)}{2i} = \frac{(4-i)}{i}$$

Multiply both numerator and denominator by $\frac{i}{i}$:

$$\frac{(4-i)}{i} = \frac{(4-i)}{i} \times \frac{i}{i} = \frac{(4i-i^2)}{i^2} = \frac{1+4i}{-1} = -1 - 4i$$

Simplifying Radical Expressions

☆ Find the prime factors of the numbers or expressions inside the radical.

☆ Use radical properties to simplify the radical expression:

$$\sqrt[n]{x^a} = x^{\frac{a}{n}}, \quad \sqrt[n]{xy} = x^{\frac{1}{n}} \times y^{\frac{1}{n}}, \quad \sqrt[n]{\frac{x}{y}} = \frac{x^{\frac{1}{n}}}{y^{\frac{1}{n}}}, \text{ and } \sqrt[n]{x} \times \sqrt[n]{y} = \sqrt[n]{xy}$$

Examples:

Example 1. Find the square root of $\sqrt{144x^2}$.

Solution: Find the factor of the expression $144x^2$: $144 = 12 \times 12$ and $x^2 = x \times x$, now use radical rule: $\sqrt[n]{a^n} = a$, Then: $\sqrt{12^2} = 12$ and $\sqrt{x^2} = x$
Finally: $\sqrt{144x^2} = \sqrt{12^2} \times \sqrt{x^2} = 12 \times x = 12x$

Example 2. Simplify. $\sqrt{8x^3}$

Solution: First factor the expression $8x^3$: $8x^3 = 2^3 \times x \times x \times x$, we need to find perfect squares: $8x^3 = 2^2 \times 2 \times x^2 \times x = 2^2 \times x^2 \times 2x$,
Then: $\sqrt{8x^3} = \sqrt{2^2 \times x^2} \times \sqrt{2x}$
Now use radical rule: $\sqrt[n]{a^n} = a$, Then: $\sqrt{2^2 \times x^2} \times \sqrt{(2x)} = 2x \times \sqrt{2x} = 2x\sqrt{2x}$

Example 3. Simplify. $\sqrt{27a^5b^4}$

Solution: First factor the expression $27a^5b^4$: $27a^5b^4 = 3^3 \times a^5 \times b^4$, we need to find perfect squares: $27a^5b^4 = 3^2 \times 3 \times a^4 \times a \times b^4$, Then:
$\sqrt{27a^5b^4} = \sqrt{3^2 \times a^4 \times b^4} \times \sqrt{3a}$
Now use radical rule: $\sqrt[n]{a^n} = a$, Then:
$\sqrt{3^2 \times a^4 \times b^4} \times \sqrt{3a} = 3 \times a^2 \times b^2 \times \sqrt{3a} = 3a^2b^2\sqrt{3a}$

Example 4. Write this radical in exponential form. $\sqrt[3]{x^5}$

Solution: To write a radical in exponential form, use this rule: $\sqrt[n]{x^a} = x^{\frac{a}{n}}$
Then: $\sqrt[3]{x^5} = x^{\frac{5}{3}}$

Adding and Subtracting Radical Expressions

☆ Only numbers and expressions that have the same radical part can be added or subtracted.

☆ Remember, combining "unlike" radical terms is not possible.

☆ For numbers with the same radical part, just add or subtract factors outside the radicals.

Examples:

Example 1. Simplify: $8\sqrt{2} + 4\sqrt{2}$

Solution: Since we have the same radical parts, then we can add these two radicals: Add like terms: $8\sqrt{2} + 4\sqrt{2} = 12\sqrt{2}$

Example 2. Simplify: $2\sqrt{8} - 2\sqrt{2}$

Solution: The two radical parts are not the same. First, we need to simplify the $2\sqrt{8}$. Then: $2\sqrt{8} = 2\sqrt{4 \times 2} = 2(\sqrt{4})(\sqrt{2}) = 4\sqrt{2}$
Now, combine like terms: $2\sqrt{8} - 2\sqrt{2} = 4\sqrt{2} - 2\sqrt{2} = 2\sqrt{2}$

Example 3. Simplify: $5\sqrt{27} + 3\sqrt{3}$

Solution: The two radical parts are not the same. First, we need to simplify the $5\sqrt{27}$. Then: $5\sqrt{27} = 5\sqrt{9 \times 3} = 5(\sqrt{9})(\sqrt{3}) = 15\sqrt{3}$
Now, add: $5\sqrt{27} + 3\sqrt{3} = 15\sqrt{3} + 3\sqrt{3} = 18\sqrt{3}$

Example 4. Simplify: $5\sqrt{8} - 2\sqrt{2}$

Solution: The two radical parts are not the same. First, we need to simplify the $5\sqrt{8}$. Then: $5\sqrt{8} = 5\sqrt{4 \times 2} = 5(\sqrt{4})(\sqrt{2}) = 10\sqrt{2}$
Now, combine like terms: $5\sqrt{8} - 2\sqrt{2} = 10\sqrt{2} - 2\sqrt{2} = 8\sqrt{2}$

Multiplying Radical Expressions

To multiply radical expressions:

☆ Multiply the numbers and expressions outside of the radicals.

☆ Multiply the numbers and expressions inside the radicals.

☆ Simplify if needed.

Examples:

Example 1. *Evaluate.* $2\sqrt{5} \times \sqrt{3}$

Solution: Multiply the numbers outside of the radicals and the radical parts.
Then: $2\sqrt{5} \times \sqrt{3} = 2 \times 1 \times \sqrt{5} \times \sqrt{3} = 2\sqrt{15}$

Example 2. *Multiply.* $3x\sqrt{3} \times 4\sqrt{x}$

Solution: Multiply the numbers outside of the radicals and the radical parts.
Then, simplify: $3x\sqrt{3} \times 4\sqrt{x} = (3x \times 4) \times \left(\sqrt{3} \times \sqrt{x}\right) = (12x)\left(\sqrt{3x}\right) = 12x\sqrt{3x}$

Example 3. *Evaluate.* $5a\sqrt{5b} \times \sqrt{2b}$

Solution: Multiply the numbers outside of the radicals and the radical parts.
Then: $5a\sqrt{5b} \times \sqrt{2b} = 5a \times 1 \times \sqrt{5b} \times \sqrt{2b} = 5a\sqrt{10b^2}$
Simplify: $5a\sqrt{10b^2} = 5a \times \sqrt{10} \times \sqrt{b^2} = 5ab\sqrt{10}$

Example 4. Evaluate. $6a\sqrt{7b} \times 3\sqrt{2b}$

Solution: Multiply the numbers outside of the radicals and the radical parts.
Then: $6a\sqrt{7b} \times 3\sqrt{2b} = 6a \times 3 \times \sqrt{7b} \times \sqrt{2b} = 18a\sqrt{14b^2}$
Simplify: $18a\sqrt{14b^2} = 18a \times \sqrt{14} \times \sqrt{b^2} = 18ab\sqrt{14}$

Rationalizing Radical Expressions

☆ Radical expressions cannot be in the denominator. (number in the bottom)

☆ To get rid of the radical in the denominator, multiply both numerator and denominator by the radical in the denominator.

☆ If there is a radical and another integer in the denominator, multiply both numerator and denominator by the conjugate of the denominator.

☆ The conjugate of $(a + b)$ is $(a - b)$ and vice versa.

Examples:

Example 1. Simplify $\frac{6}{\sqrt{2}}$

Solution: Multiply both numerator and denominator by $\sqrt{2}$. Then:

$\frac{6}{\sqrt{2}} \times \frac{\sqrt{2}}{\sqrt{2}} = \frac{6\sqrt{2}}{\sqrt{4}} = \frac{6\sqrt{2}}{2}$, Now, simplify: $\frac{6\sqrt{2}}{2} = 3\sqrt{2}$

Example 2. Simplify $\frac{5}{\sqrt{6}-4}$

Solution: Multiply by the conjugate: $\frac{\sqrt{6}+4}{\sqrt{6}+4} \rightarrow \frac{5}{\sqrt{6}-4} \times \frac{\sqrt{6}+4}{\sqrt{6}+4}$

$\left(\sqrt{6}-4\right)\left(\sqrt{6}+4\right) = -10$, then: $\frac{5}{\sqrt{6}-4} \times \frac{\sqrt{6}+4}{\sqrt{6}+4} = \frac{5(\sqrt{6}+4)}{-10}$

Use the fraction rule: $\frac{a}{-b} = -\frac{a}{b} \rightarrow \frac{5(\sqrt{6}+4)}{-10} = -\frac{5(\sqrt{6}+4)}{10} = -\frac{1}{2}(\sqrt{6}+4)$

Example 3. Simplify $\frac{2}{\sqrt{3}-1}$

Solution: Multiply by the conjugate: $\frac{\sqrt{3}+1}{\sqrt{3}+1}$

$\frac{2}{\sqrt{3}-1} \times \frac{\sqrt{3}+1}{\sqrt{3}+1} = \frac{2(\sqrt{3}+1)}{2} \rightarrow = (\sqrt{3}+1)$

Radical Equations

To solve a radical equation:

☆ Isolate the radical on one side of the equation.

☆ Square both sides of the equation to remove the radical.

☆ Solve the equation for the variable.

☆ Plug in the answer (answers) into the original equation to avoid extraneous values.

Examples:

Example 1. Solve $\sqrt{x} - 5 = 15$

Solution: Add 5 to both sides: $\sqrt{x} = 20$

Square both sides: $\left(\sqrt{x}\right)^2 = 20^2 \rightarrow x = 400$

Plug in the value of 400 for x in the original equation and check the answer:

$x = 400 \rightarrow \sqrt{x} - 5 = \sqrt{400} - 5 = 20 - 5 = 15$

So, the value of 400 for x is correct.

Example 2. What is the value of x in this equation?
$$2\sqrt{x+1} = 4$$

Solution: Divide both sides by 2. Then:

$2\sqrt{x+1} = 4 \rightarrow \frac{2\sqrt{x+1}}{2} = \frac{4}{2} \rightarrow \sqrt{x+1} = 2$

Square both sides: $\left(\sqrt{(x+1)}\right)^2 = 2^2$, Then: $x + 1 = 4 \rightarrow x = 3$

Substitute x by 3 in the original equation and check the answer:

$x = 3 \rightarrow 2\sqrt{x+1} = 2\sqrt{3+1} = 2\sqrt{4} = 2(2) = 4$

So, the value of 3 for x is correct.

Domain and Range of Radical Functions

☆ To find the domain of a radic al function, find all possible values of the variable inside radical.

☆ Remember that having a negative number under the square root symbol is not possible. (For cubic roots, we can have negative numbers)

☆ To find the range, plug in the minimum and maximum values of the variable inside radical.

Example:

Example 1. Find the domain and range of the radical function. $y = \sqrt{x - 3}$

Solution: For domain: Find non-negative values for radicals: $x - 3 \geq 0$

Domain of functions: $x - 3 \geq 0 \rightarrow x \geq 3$

Domain of the function $y = \sqrt{x - 3}$: $x \geq 3$

For range: The range of a radical function of the form $c\sqrt{ax + b} + k$ is: $f(x) \geq k$

For the function $y = \sqrt{x - 3}$, the value of k is 0. Then: $f(x) \geq 0$

Range of the function $y = \sqrt{x - 3}$: $f(x) \geq 0$

Example 2. Find the domain and range of the radical function. $y = 5\sqrt{3x + 6} + 4$

Solution: For domain: Find non-negative values for radicals: $3x + 6 \geq 0$

Domain of functions: $3x + 6 \geq 0 \rightarrow 3x \geq -6 \rightarrow x \geq -2$

Domain of the function $y = 5\sqrt{3x + 6} + 4$: $x \geq -2$

For range: The range of a radical function of the form $c\sqrt{ax + b} + k$ is: $f(x) \geq k$

For the function $y = 5\sqrt{3x + 6} + 4$, the value of k is 4. Then: $f(x) \geq 4$

Range of the function $y = 5\sqrt{3x + 6} + 4$: $f(x) \geq 4$

Day 9: Practices

✎ *Evaluate.*

1) $(-5i) - (7i) =$

2) $(-2i) + (-8i) =$

3) $(2i) - (6 + 3i) =$

4) $(4 - 6i) + (-2i) =$

5) $(-7i) + (4 + 5i) =$

6) $10 + (-2 - 6i) =$

7) $(-3i) - (9 + 2i) =$

8) $(4 + 6i) - (-3i) =$

✎ *Calculate.*

9) $(3 - 2i)(4 - 3i) =$

10) $(6 + 2i)(3 + 2i) =$

11) $(8 - i)(4 - 2i) =$

12) $(2 - 4i)(3 - 5i) =$

13) $(5 + 6i)(3 + 2i) =$

14) $(5 + 3i)(9 + 2i) =$

✎ *Simplify.*

15) $\frac{3}{2i} =$

16) $\frac{8}{-3i} =$

17) $\frac{-9}{2i} =$

18) $\frac{2-3i}{-5i} =$

19) $\frac{4-5i}{-2i} =$

20) $\frac{8+3i}{2i} =$

✎ *Simplify.*

21) $\sqrt{256y} =$

22) $\sqrt{900} =$

23) $\sqrt{144a^2b} =$

24) $\sqrt{36 \times 9} =$

✐ *Simplify.*

25) $3\sqrt{5} + 2\sqrt{5} =$

26) $6\sqrt{3} + 4\sqrt{27} =$

27) $5\sqrt{2} + 10\sqrt{18} =$

28) $7\sqrt{2} - 5\sqrt{8} =$

✐ *Evaluate.*

29) $\sqrt{5} \times \sqrt{3} =$

30) $\sqrt{6} \times \sqrt{8} =$

31) $3\sqrt{5} \times \sqrt{9} =$

32) $2\sqrt{3} \times 3\sqrt{7} =$

✐ *Simplify.*

33) $\frac{1}{\sqrt{3}-6} =$

34) $\frac{5}{\sqrt{2}+7} =$

35) $\frac{\sqrt{3}}{1-\sqrt{6}} =$

36) $\frac{2}{\sqrt{3}+5} =$

✐ *Solve for x.*

37) $\sqrt{x} + 2 = 9$

38) $3 + \sqrt{x} = 12$

39) $\sqrt{x} + 5 = 30$

40) $\sqrt{x} - 9 = 27$

41) $10 = \sqrt{x+1}$

42) $\sqrt{x+4} = 3$

✐ *Identify the domain and range of each function.*

43) $y = \sqrt{x+2} - 1$

44) $y = \sqrt{x+1}$

45) $y = \sqrt{x-4}$

46) $y = \sqrt{x-3} + 1$

Day 9: Answers

1) $-12i$(Remove parentheses by multiplying -1 to the second parentheses: $(-5i) - (7i) = -5i - 7i$.Combine like terms: $-5i - 7i = -12i$)

2) $-10i$(Remove parentheses: $(-2i) + (-8i) = -2i - 8i$. Combine like terms: $-2i - 8i = -10i$)

3) $-6 - i$(Remove parentheses by multiplying -1 to the second parentheses: $(2i) - (6 + 3i) = 2i - 6 - 3i$.Combine like terms: $2i - 6 - 3i = -6 - i$)

4) $4 - 8i$(Remove parentheses: $(4 - 6i) + (-2i) = 4 - 6i - 2i$.Combine like terms: $4 - 6i - 2i = 4 - 8i$)

5) $4 - 2i$(Remove parentheses: $(-7i) + (4 + 5i) = -7i + 4 + 5i$.Combine like terms: $-7i + 4 + 5i = 4 - 2i$)

6) $8 - 6i$(Remove parentheses: $10 + (-2 - 6i) = 10 - 2 - 6i$.Combine like terms: $10 - 2 - 6i = 8 - 6i$)

7) $-9 - 5i$(Remove parentheses by multiplying -1 to the second parentheses: $(-3i) - (9 + 2i) = -3i - 9 - 2i$.Combine like terms: $-3i - 9 - 2i = -9 - 5i$)

8) $4 + 9i$(Remove parentheses by multiplying -1 to the second parentheses: $(4 + 6i) - (-3i) = 4 + 6i + 3i$.Combine like terms: $4 + 6i + 3i = 4 + 9i$)

9) $6 - 17i$ (Use the multiplication of imaginary numbers rule: $(a + bi) + (c + di) = (ac - bd) + (ad + bc)i \rightarrow \big(3 \times 4 - (-2)(-3)\big) + (3(-3) + (-2) \times 4)i = 6 - 17i$)

10) $14 + 18i$ (Use the multiplication of imaginary numbers rule: $(a + bi) + (c + di) = (ac - bd) + (ad + bc)i \rightarrow (3 \times 6 - (2 \times 2)) + (6 \times 2 + 2 \times 3)i = 14 + 18i$)

11) $30 - 20i$ (Use the multiplication of imaginary numbers rule: $(a + bi) + (c + di) = (ac - bd) + (ad + bc)i \rightarrow \big(8 \times 4 - (-1)(-2)\big) + (8(-2) + (-1) \times 4)i = 30 - 20i$)

12) $-14 - 22i$ (Use the multiplication of imaginary numbers rule: $(a + bi) + (c + di) = (ac - bd) + (ad + bc)i \rightarrow \big(2 \times 3 - (-4)(-5)\big) + (2(-5) + (-4) \times 3)i = -14 - 22i$)

13) $3 + 28i$ (Use the multiplication of imaginary numbers rule: $(a + bi) + (c + di) = (ac - bd) + (ad + bc)i \rightarrow (3 \times 5 - (6 \times 2)) + (5 \times 2 + 6 \times 3)i = 3 + 28i$)

14) $39 + 37i$ (Use the multiplication of imaginary numbers rule: $(a + bi) + (c + di) = (ac - bd) + (ad + bc)i \rightarrow (5 \times 9 - (3 \times 2)) + (5 \times 2 + 3 \times 9)i = 39 + 37i$)

15) $-\frac{3i}{2}$ (Multiply both numerator and denominator by $\frac{i}{i}$: $\frac{3}{2i} = \frac{3 \times (i)}{2i(i)} = \frac{3 \times (i)}{2(i^2)} = \frac{3i}{2(-1)} = \frac{3i}{-2} = -\frac{3i}{2}$)

16) $\frac{8i}{3}$ (Multiply both numerator and denominator by $\frac{i}{i}$: $\frac{8}{-3i} = \frac{8 \times (i)}{-3i(i)} = \frac{8 \times (i)}{-3(i^2)} = \frac{8i}{-3(-1)} = \frac{8i}{3}$)

17) $\frac{9i}{2}$ (Multiply both numerator and denominator by $\frac{i}{i}$: $\frac{-9}{2i} = \frac{-9 \times (i)}{2i(i)} = \frac{-9 \times (i)}{2(i^2)} = \frac{-9i}{2(-1)} = \frac{-9i}{-2} = \frac{9i}{2}$)

18) $\frac{3}{5} + \frac{2}{5}i$ (Multiply both numerator and denominator by $\frac{i}{i}$: $\frac{2-3i}{-5i} = \frac{(2-3i)(i)}{-5i(i)} = \frac{(2)(i)-(3i)(i)}{-5(i^2)} = \frac{2i-3i^2}{-5(-1)} = \frac{2i-3(-1)}{5} = \frac{2i}{5} + \frac{3}{5} = \frac{3}{5} + \frac{2}{5}i$)

19) $\frac{5}{2} + 2i$ (Multiply both numerator and denominator by $\frac{i}{i}$: $\frac{4-5i}{-2i} = \frac{(4-5i)(i)}{-2i(i)} = \frac{(4)(i)-(5i)(i)}{-2(i^2)} = \frac{4i-5i^2}{-2(-1)} = \frac{4i-5(-1)}{2} = \frac{4i}{2} + \frac{5}{2} = \frac{5}{2} + 2i$)

$\frac{3}{2} - 4i$ (Multiply both numerator and denominator by $\frac{i}{i}$: $\frac{8+3i}{2i} = \frac{(8+3i)(i)}{2i(i)} = \frac{(8)(i)+(3i)(i)}{2(i^2)} = \frac{8i+3i^2}{2(-1)} = \frac{8i+3(-1)}{-2} = -\frac{8i}{2} + \frac{3}{2} = \frac{3}{2} - 4i$)

20) $16\sqrt{y}$ (First factor the expression $256y$: $256y = 16 \times 16 \times y$. Now use radical rule: $\sqrt[n]{a^n} = a$, Then: $\sqrt{16^2} \times \sqrt{y} = 16 \times \sqrt{y} = 16\sqrt{y}$)

21) 30 (Find the factor of the expression 900: $900 = 9 \times 100 \rightarrow 9 = 3 \times 3$ and $100 = 10 \times 10$, now use radical rule: $\sqrt[n]{a^n} = a$, Then: $\sqrt{3^2} = 3$ and $\sqrt{10^2} = 10$. Finally: $\sqrt{900} = \sqrt{3^2} \times \sqrt{10^2} = 3 \times 10 = 30$)

22) $12a\sqrt{b}$ (First factor the expression $144a^2b$: $144a^2b = 12^2 \times a^2 \times b$, Then: $\sqrt{144a^2b} = \sqrt{12^2 \times a^2} \times \sqrt{b}$. Now use radical rule: $\sqrt[n]{a^n} = a$, Then: $\sqrt{12^2 \times a^2} \times \sqrt{b} = 12 \times a \times \sqrt{b} = 12a\sqrt{b}$)

23) 18 (Find the factor of the expression 36×9: $36 \times 9 \rightarrow 36 = 6 \times 6$ and $9 = 3 \times 3$, now use radical rule: $\sqrt[n]{a^n} = a$, Then: $\sqrt{6^2} = 6$ and $\sqrt{3^2} = 3$. Finally: $\sqrt{36 \times 9} = \sqrt{6^2} \times \sqrt{3^2} = 6 \times 3 = 18$)

24) $5\sqrt{5}$ (Since we have the same radical parts, then we can add these two radicals: Add like terms: $3\sqrt{5} + 2\sqrt{5} = 5\sqrt{5}$)

25) $18\sqrt{3}$ (The two radical parts are not the same. First, we need to simplify the $4\sqrt{27}$. Then: $4\sqrt{27} = 4\sqrt{9 \times 3} = 4(\sqrt{9})(\sqrt{3}) = 12\sqrt{3}$. Now, combine like terms: $6\sqrt{3} + 4\sqrt{27} = 6\sqrt{3} + 12\sqrt{3} = 18\sqrt{3}$)

26) $35\sqrt{2}$ (The two radical parts are not the same. First, we need to simplify the $10\sqrt{18}$. Then: $10\sqrt{18} = 10\sqrt{9 \times 2} = 10(\sqrt{9})(\sqrt{2}) = 30\sqrt{2}$. Now, combine like terms: $5\sqrt{2} + 10\sqrt{18} = 5\sqrt{2} + 30\sqrt{2} = 35\sqrt{2}$)

27) $-3\sqrt{2}$ (The two radical parts are not the same. First, we need to simplify the $5\sqrt{8}$. Then: $5\sqrt{8} = 5\sqrt{4 \times 2} = 5(\sqrt{4})(\sqrt{2}) = 10\sqrt{2}$. Now, combine like terms: $7\sqrt{2} - 5\sqrt{8} = 7\sqrt{2} - 10\sqrt{2} = -3\sqrt{2}$)

28) $\sqrt{15}$ (Multiply the numbers outside of the radicals and the radical parts. Then: $\sqrt{5} \times \sqrt{3} = \sqrt{5} \times \sqrt{3} = \sqrt{15}$)

29) $4\sqrt{3}$ (Multiply the numbers outside of the radicals and the radical parts. Then, simplify: $\sqrt{6} \times \sqrt{8} = \sqrt{2 \times 3} \times \sqrt{2^3} = \sqrt{2^4 \times 3} \rightarrow 4\sqrt{3} =$, then: $\sqrt{6} \times \sqrt{8} = 4\sqrt{3}$)

30) $9\sqrt{5}$ (Multiply the numbers outside of the radicals and the radical parts. Then, simplify: $3\sqrt{5} \times \sqrt{9} = (3) \times (\sqrt{5} \times \sqrt{9}) = (3)(\sqrt{45}) = 3\sqrt{45} \rightarrow \sqrt{45} = \sqrt{5 \times 3^2} = 3\sqrt{5}$, then: $3\sqrt{45} = 3 \times 3\sqrt{5} = 9\sqrt{5}$)

31) $6\sqrt{21}$ (Multiply the numbers outside of the radicals and the radical parts. Then, simplify: $2\sqrt{3} \times 3\sqrt{7} = (2 \times 3) \times (\sqrt{3} \times \sqrt{7}) = (6)(\sqrt{21}) = 6\sqrt{21}$, then: $2\sqrt{3} \times 3\sqrt{7} = 6\sqrt{21}$)

32) $-\frac{\sqrt{3}+6}{33}$ (Multiply by the conjugate: $\frac{\sqrt{3}+6}{\sqrt{3}+6} \rightarrow \frac{1}{\sqrt{3}-6} \times \frac{\sqrt{3}+6}{\sqrt{3}+6} \rightarrow (\sqrt{3}-6)(\sqrt{3}+$

6$) = -33$, then: $\frac{1}{\sqrt{3}-6} \times \frac{\sqrt{3}+6}{\sqrt{3}+6} = \frac{1(\sqrt{3}+6)}{-33}$. Use the fraction rule: $\frac{a}{-b} = -\frac{a}{b} \rightarrow$

$\frac{1(\sqrt{3}+6)}{-33} = -\frac{1(\sqrt{3}+6)}{33} = -\frac{\sqrt{3}+6}{33}$)

33) $-\frac{5(\sqrt{2}-7)}{47}$ (Multiply by the conjugate: $\frac{\sqrt{2}-7}{\sqrt{2}-7} \rightarrow \frac{5}{\sqrt{2}+7} \times \frac{\sqrt{2}-7}{\sqrt{2}-7} \rightarrow (\sqrt{2}-7)(\sqrt{2}+$

7$) = -47$, then: $\frac{5}{\sqrt{2}+7} \times \frac{\sqrt{2}-7}{\sqrt{2}-7} = \frac{5(\sqrt{2}-7)}{-47}$. Use the fraction rule: $\frac{a}{-b} = -\frac{a}{b} \rightarrow$

$\frac{5(\sqrt{2}-7)}{-47} = -\frac{5(\sqrt{2}-7)}{47} = -\frac{5(\sqrt{2}-7)}{47}$)

34) $-\frac{\sqrt{3}+3\sqrt{2}}{5}$ (Multiply by the conjugate: $\frac{1+\sqrt{6}}{1+\sqrt{6}} \rightarrow \frac{\sqrt{3}}{1-\sqrt{6}} \times \frac{1+\sqrt{6}}{1+\sqrt{6}} \rightarrow (1+\sqrt{6})(1-\sqrt{6}) =$

-5, then: $\frac{\sqrt{3}}{1-\sqrt{6}} \times \frac{1+\sqrt{6}}{1+\sqrt{6}} = \frac{\sqrt{3}(1+\sqrt{6})}{-5}$. Use the fraction rule: $\frac{a}{-b} = -\frac{a}{b} \rightarrow \frac{\sqrt{3}(1+\sqrt{6})}{-5} =$

$-\frac{\sqrt{3}(1+\sqrt{6})}{5} = -\frac{\sqrt{3}+\sqrt{18}}{5} = -\frac{\sqrt{3}+3\sqrt{2}}{5}$)

35) $-\frac{\sqrt{3}-5}{11}$ (Multiply by the conjugate: $\frac{\sqrt{3}-5}{\sqrt{3}-5} \rightarrow \frac{2}{\sqrt{3}+5} \times \frac{\sqrt{3}-5}{\sqrt{3}-5} \rightarrow (\sqrt{3}-5)(\sqrt{3}+5) =$

-22, then: $\frac{2}{\sqrt{3}+5} \times \frac{\sqrt{3}-5}{\sqrt{3}-5} = \frac{2(\sqrt{3}-5)}{-22}$. Use the fraction rule: $\frac{a}{-b} = -\frac{a}{b} \rightarrow \frac{2(\sqrt{3}-5)}{-22} =$

$-\frac{2(\sqrt{3}-5)}{22} = -\frac{\sqrt{3}-5}{11}$)

36) 49 (Subtract 2 from both sides: $\sqrt{x} = 7$. Square both sides: $\left(\sqrt{x}\right)^2 = 7^2 \rightarrow x =$ 49. Plug in the value of 49 for x in the original equation and check the answer: $x = 49 \rightarrow \sqrt{x} + 2 = \sqrt{49} + 2 = 7 + 2 = 9$)

37) 81 (Subtract 3 from both sides: $\sqrt{x} = 9$. Square both sides: $\left(\sqrt{x}\right)^2 = 9^2 \rightarrow x =$ 81. Plugin the value of 81 for x in the original equation and check the answer: $x = 81 \rightarrow 3 + \sqrt{x} = 3 + \sqrt{81} = 3 + 9 = 12$)

38) 625 (Subtract 5 from both sides: $\sqrt{x} = 25$. Square both sides: $\left(\sqrt{x}\right)^2 = 25^2 \rightarrow$ $x = 625$. Plug in the value of 625 for x in the original equation and check the answer: $x = 625 \rightarrow \sqrt{x} + 5 = \sqrt{625} + 5 = 25 + 5 = 30$)

39) 1,296 (Add 9 to both sides: $\sqrt{x} = 36$. Square both sides: $\left(\sqrt{x}\right)^2 = 36^2 \rightarrow x =$ 1,296. Plug in the value of 1,296 for x in the original equation and check the answer: $x = 1,296 \rightarrow \sqrt{x} - 9 = \sqrt{1,296} - 9 = 36 - 9 = 27$)

40) 99 (Square both sides: $10^2 = \left(\sqrt{(x+1)}\right)^2$, Then: $100 = x + 1 \rightarrow 100 \rightarrow x = 99$. Substitute x by 99 in the original equation and check the answer: $x = 99 \rightarrow \sqrt{x+1} = \sqrt{99+1} = \sqrt{100} = 10$. So, the value of 99 for x is correct.)

41) 5 (Square both sides: $\left(\sqrt{(x+4)}\right)^2 = 3^2$, Then: $x + 4 = 9 \rightarrow x = 5$. Substitute x by 5 in the original equation and check the answer: $x = 5 \rightarrow \sqrt{x+4} = \sqrt{5+4} = \sqrt{9} = 3$. So, the value of 5 for x is correct.)

42) $x \geq -2, y \geq -1$ (For domain: Find non-negative values for radicals: $x + 2 \geq 0$. Domain of functions: $x + 2 \geq 0 \rightarrow x \geq -2$ .Domain of the function $y = \sqrt{x+2} - 1$: $x \geq -2$. For range: The range of a radical function of the form $c\sqrt{ax+b} + k$ is: $f(x) \geq k$. For the function $y = \sqrt{x+2} - 1$, the value of k is -1. Then: $f(x) \geq -1$.Range of the function $y = \sqrt{x+2} - 1$: $f(x) \geq -1$)

43) $x \geq -1, y \geq 0$ (For domain: Find non-negative values for radicals: $x + 1 \geq 0$. Domain of functions: $x + 1 \geq 0 \rightarrow x \geq -1$ .Domain of the function $y = \sqrt{x+1}$: $x \geq -1$. For range: The range of a radical function of the form $c\sqrt{ax+b} + k$ is: $f(x) \geq k$. For the function $y = \sqrt{x+1}$, the value of k is 0. Then: $f(x) \geq 0$.Range of the function $y = \sqrt{x+1}$: $f(x) \geq 0$)

44) $x \geq 4, y \geq 0$ (For domain: Find non-negative values for radicals: $x - 4 \geq 0$. Domain of functions: $x - 4 \geq 0 \rightarrow x \geq 4$ .Domain of the function $y = \sqrt{x-4}$: $x \geq 4$. For range: The range of a radical function of the form $c\sqrt{ax+b} + k$ is: $f(x) \geq k$. For the function $y = \sqrt{x-4}$, the value of k is 0. Then: $f(x) \geq 0$.Range of the function $y = \sqrt{x-4}$: $f(x) \geq 0$)

45) $x \geq 3, y \geq 1$ (For domain: Find non-negative values for radicals:

$x - 3 \geq 0$. Domain of functions: $x - 3 \geq 0 \rightarrow x \geq 3$ .Domain of the function $y = \sqrt{x-3}$: $x \geq 3$. For range: The range of a radical function of the form $c\sqrt{ax+b} + k$ is: $f(x) \geq k$. For the function $y = \sqrt{x-3} + 1$, the value of k is 1. Then: $f(x) \geq 1$. Range of the function $y = \sqrt{x-3} + 1$: $f(x) \geq 1$)

DAY 10 Rational Expressions and Trigonometric Functions

Math topics that you'll learn in this chapter:

133

Simplifying Complex Fractions

✩ Convert mixed numbers to improper fractions.

✩ Simplify all fractions.

✩ Write the fraction in the numerator of the main fraction line then write division sign (÷) and the fraction of the denominator.

✩ Use normal method for dividing fractions.

✩ Simplify as needed.

Example:

Example 1. Simplify: $\dfrac{\frac{2}{5} \div \frac{1}{3}}{\frac{5}{9} + \frac{1}{3}}$

Solution: First, simplify the numerator: $\dfrac{2}{5} \div \dfrac{1}{3} = \dfrac{6}{5}$, then, simplify the denominator: $\dfrac{5}{9} + \dfrac{1}{3} = \dfrac{8}{9}$, Now, write the complex fraction using the division sign (÷): $\dfrac{\frac{2}{5} \div \frac{1}{3}}{\frac{5}{9} + \frac{1}{3}} = \dfrac{\frac{6}{5}}{\frac{8}{9}} = \dfrac{6}{5} \div \dfrac{8}{9}$, Use the dividing fractions rule: (Keep, Change, Flip) $\dfrac{6}{5} \div \dfrac{8}{9} = \dfrac{6}{5} \times \dfrac{9}{8} = \dfrac{54}{40} = \dfrac{27}{20} = 1\dfrac{7}{20}$

Example 2. Simplify: $\dfrac{\frac{3}{4}}{\frac{3}{22} - \frac{5}{18}}$

Solution: First, simplify the denominator: $\dfrac{3}{22} - \dfrac{5}{18} = -\dfrac{14}{99}$,

Then: $\dfrac{\frac{3}{4}}{\frac{3}{22} - \frac{5}{18}} = \dfrac{\frac{3}{4}}{-\frac{14}{99}}$; Now, write the complex fraction using the division sign: $\dfrac{\frac{3}{4}}{-\frac{14}{99}} = \dfrac{3}{4} \div \left(-\dfrac{14}{99}\right)$. Use the dividing fractions rule: Keep, Change, Flip (keep the first fraction, change the division sign to multiplication, flip the second fraction) $\dfrac{3}{4} \div \left(-\dfrac{14}{99}\right) = \dfrac{3}{4} \times \left(-\dfrac{99}{14}\right) = -\dfrac{297}{56}$

Graphing Rational Expressions

☆ A rational expression is a fraction in which the numerator and/or the denominator are polynomials. Examples: $\frac{1}{x}, \frac{x^2}{x-1}, \frac{x^2-x+2}{x^2+5x+1}, \frac{m^2+6m-5}{m-2m}$

☆ To graph a rational function:

- Find the vertical asymptotes of the function if there is any. (Vertical asymptotes are vertical lines which correspond to the zeroes of the denominator. The graph will have a vertical asymptote at $x = a$ if the denominator is zero at $x = a$ and the numerator isn't zero at $x = a$)

- Find the horizontal or slant asymptote. (If the numerator has a bigger degree than the denominator, there will be a slant asymptote. To find the slant asymptote, divide the numerator by the denominator using either long division or synthetic division.)

- If the denominator has a bigger degree than the numerator, the horizontal asymptote is the x-axes or the line $y = 0$. If they have the same degree, the horizontal asymptote equals the leading coefficient (the coefficient of the largest exponent) of the numerator divided by the leading coefficient of the denominator.

- Find intercepts and plug in some values of x and solve for y, then graph the function.

Example:

Example 1. Graph rational function. $f(x) = \frac{x^2-x+2}{x-1}$

Solution: First, notice that the graph is in two pieces. Most rational functions have graphs in multiple pieces. Find $y - intercept$ by substituting zero for x and solving for y $(f(x))$: $x = 0 \rightarrow y = \frac{x^2-x+2}{x-1} = \frac{0^2-0+2}{0-1} = -2$,

$$y - intercept: (0, -2)$$

Asymptotes of $\frac{x^2-x+2}{x-1}$: Vertical: $x = 1$, Slant asymptote: $y = 2x + 1$ (divide the numerator by the denominator). After finding the asymptotes, you can plug in some values for x and solve for y. Here is the sketch for this function.

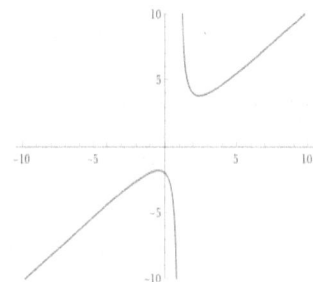

Adding and Subtracting Rational Expressions

For adding and subtracting rational expressions:

☆ Find least common denominator (LCD).

☆ Write each expression using the LCD.

☆ Add or subtract the numerators.

☆ Simplify as needed.

Examples:

Example 1. Solve. $\frac{4}{2x+3}+\frac{x-2}{2x+3}=$

Solution: The denominators are equal. Then, use fractions addition rule:

$\frac{a}{c}\pm\frac{b}{c}=\frac{a\pm b}{c}\rightarrow\frac{4}{2x+3}+\frac{x-2}{2x+3}=\frac{4+(x-2)}{2x+3}=\frac{x+2}{2x+3}$

Example 2. Solve. $\frac{x+4}{x-5}+\frac{x-4}{x+6}=$

Solution: Find the least common denominator of $(x-5)$ and $(x+6)$: $(x-5)(x+6)$

Then: $\frac{x+4}{x-5}+\frac{x-4}{x+6}=\frac{(x+4)(x+6)}{(x-5)(x+6)}+\frac{(x-4)(x-5)}{(x+6)(x-5)}=\frac{(x+4)(x+6)+(x-4)(x-5)}{(x+6)(x-5)}$

Expand: $(x+4)(x+6)+(x-4)(x-5)=2x^2+x+44$

Then: $\frac{(x+4)(x+6)+(x-4)(x-5)}{(x+6)(x-5)}=\frac{2x^2+x+44}{(x+6)(x-5)}=\frac{2x^2+x+44}{x^2+x-30}$

Example 3. Solve. $\frac{x+2}{x-4}+\frac{x-2}{x+5}=$

Solution: Find the least common denominator of $(x-5)$ and $(x+6)$: $(x-5)(x+6)$

Then: $\frac{x+2}{x-4}+\frac{x-2}{x+5}=\frac{(x+2)(x+5)}{(x-4)(x+5)}+\frac{(x-2)(x-4)}{(x-4)(x+5)}=\frac{(x+2)(x+5)+(x-2)(x-4)}{(x-4)(x+5)}$

Expand: $(x+2)(x+5)+(x-2)(x-4)=2x^2+x+18$

Then: $\frac{(x+2)(x+5)+(x-2)(x-4)}{(x-4)(x+5)}=\frac{2x^2+x+18}{(x-4)(x+5)}=\frac{2x^2+x+18}{x^2+x-20}$

Multiplying Rational Expressions

☆ Multiplying rational expressions is the same as multiplying fractions. First, multiply numerators and then multiply denominators. Then, simplify as needed.

Examples:

Example 1. Solve: $\frac{x+6}{x-1} \times \frac{x-1}{5} =$

Solution: Multiply numerators and denominators: $\frac{a}{b} \times \frac{c}{d} = \frac{a \times c}{b \times d}$

$\frac{x+6}{x-1} \times \frac{x-1}{5} = \frac{(x+6)(x-1)}{5(x-1)}$

Cancel the common factor: $(x-1)$

Then: $\frac{(x+6)(x-1)}{5(x-1)} = \frac{(x+6)}{5}$

Example 2. Solve: $\frac{x-2}{x+3} \times \frac{2x+6}{x-2} =$

Solution: Multiply numerators and denominators: $\frac{x-2}{x+3} \times \frac{2x+6}{x-2} = \frac{(x-2)(2x+6)}{(x+3)(x-2)}$

Cancel the common factor: $\frac{(x-2)(2x+6)}{(x+3)(x-2)} = \frac{(2x+6)}{(x+3)}$

Factor $2x + 6 = 2(x+3)$, Then: $\frac{2(x+3)}{(x+3)} = 2$

Example 3. Solve: $\frac{x+3}{x-2} \times \frac{x-2}{4} =$

Solution: Multiply numerators and denominators: $\frac{a}{b} \times \frac{c}{d} = \frac{a \times c}{b \times d}$

$\frac{x+3}{x-2} \times \frac{x-2}{4} = \frac{(x+3)(x-2)}{4(x-2)}$, Cancel the common factor: $(x-2)$

Then: $\frac{(x+3)(x-2)}{4(x-2)} = \frac{(x+3)}{4}$

Dividing Rational Expressions

☆ To divide rational expressions, use the same method we use for dividing fractions. (Keep, Change, Flip)

☆ Keep the first rational expression, change the division sign to multiplication, and flip the numerator and denominator of the second rational expression. Then, multiply numerators and multiply denominators. Simplify as needed.

Examples:

Example 1. Solve. $\frac{x+2}{3x} \div \frac{x^2+5x+6}{3x^2+3x} =$

Solution: Use fractions division rule: $\frac{a}{b} \div \frac{c}{d} = \frac{a}{b} \times \frac{d}{c} = \frac{a \times d}{b \times c}$

$$\frac{x+2}{3x} \div \frac{x^2+5x+6}{3x^2+3x} = \frac{x+2}{3x} \times \frac{3x^2+3x}{x^2+5x+6} = \frac{(x+2)(3x^2+3x)}{(3x)(x^2+5x+6)}$$

Now, factorize the expressions $3x^2 + 3x$ and $(x^2 + 5x + 6)$. Then:

$3x^2 + 3x = 3x(x + 1)$ and $x^2 + 5x + 6 = (x + 2)(x + 3)$

Simplify: $\frac{(x+2)(3x^2+3x)}{(3x)(x^2+5x+6)} = \frac{(x+2)(3x)(x+1)}{(3x)(x+2)(x+3)}$, cancel common factors. Then:

$$\frac{(x+2)(3x)(x+1)}{(3x)(x+2)(x+3)} = \frac{x+1}{x+3}$$

Example 2. Solve. $\frac{7x}{x+5} \div \frac{x}{4x+8} =$

Solution: Use fractions division rule: $\frac{a}{b} \div \frac{c}{d} = \frac{a}{b} \times \frac{d}{c} = \frac{a \times d}{b \times c}$

Then: $\frac{7x}{x+5} \div \frac{x}{4x+8} = \frac{7x}{x+5} \times \frac{4x+8}{x} = \frac{7x(4x+8)}{x(x+5)} = \frac{7x \times 2(x+4)}{x(x+5)}$

Cancel common factor: $\frac{7x \times 2(x+4)}{x(x+5)} = \frac{14x(x+4)}{x(x+5)} = \frac{14(x+4)}{(x+5)}$

Rational Equations

For solving rational equations, we can use following methods:

☆ **Converting to a common denominator:** In this method, you need to get a common denominator for both sides of the equation. Then, make numerators equal and solve for the variable.

☆ **Cross-multiplying:** This method is useful when there is only one fraction on each side of the equation. Simply multiply the first numerator by the second denominator and make the result equal to the product of the second numerator and the first denominator.

Examples:

Example 1. Solve. $\frac{x-2}{x+1} = \frac{x+4}{x-2}$

Solution: Use cross multiply method: if $\frac{a}{b} = \frac{c}{d}$, then: $a \times d = b \times c$

$\frac{x-2}{x+1} = \frac{x+4}{x-2} \rightarrow (x-2)(x-2) = (x+4)(x+1)$

Expand: $(x-2)^2 = x^2 - 4x + 4$ and $(x+4)(x+1) = x^2 + 5x + 4$,

Then: $x^2 - 4x + 4 = x^2 + 5x + 4$, Now, simplify: $x^2 - 4x = x^2 + 5x$, subtract both sides $(x^2 + 5x)$,

Then: $x^2 - 4x - (x^2 + 5x) = x^2 + 5x - (x^2 + 5x) \rightarrow -9x = 0 \rightarrow x = 0$

Example 2. Solve. $\frac{2x}{x-3} = \frac{2x+2}{2x-6}$

Solution: Multiply the numerator and denominator of the rational expression on the left by 2 to get a common denominator $(2x-6)$. $\frac{2(2x)}{2(x-3)} = \frac{4x}{2x-6}$

Now, the denominators on both side of the equation are equal. Therefore, their numerators must be equal too.

$\frac{4x}{2x-6} = \frac{2x+2}{2x-6} \rightarrow 4x = 2x + 2 \rightarrow 2x = 2 \rightarrow x = 1$

Angle and Angle Measure

☆ To convert degrees to radians, use this formula:

$$\text{Radians} = \text{Degrees} \times \frac{\pi}{180}$$

☆ To convert radians to degrees, use this formula:

$$\text{Degrees} = \text{Radians} \times \frac{180}{\pi}$$

Examples:

Example 1. Convert 160 degrees to radian.

Solution: Use this formula: $\text{Radians} = \text{Degrees} \times \frac{\pi}{180}$

$$\text{Radians} = 160 \times \frac{\pi}{180} = \frac{160\pi}{180} = \frac{8\pi}{9}$$

Example 2. Convert radian measure $\frac{3\pi}{4}$ to degree measure.

Solution: Use this formula: $\text{Degrees} = \text{Radians} \times \frac{180}{\pi}$

$$\text{Radians} = \frac{3\pi}{4} \times \frac{180}{\pi} = \frac{540\pi}{4\pi} = 135$$

Example 3. Convert 150 degrees to radian.

Solution: Use this formula: $\text{Radians} = \text{Degrees} \times \frac{\pi}{180}$

$$\text{Radians} = 150 \times \frac{\pi}{180} = \frac{150\pi}{180} = \frac{5\pi}{6}$$

Example 4. Convert radian measure $\frac{\pi}{4}$ to degree measure.

Solution: Use this formula: $\text{Degrees} = \text{Radians} \times \frac{180}{\pi}$

$$\text{Radians} = \frac{\pi}{4} \times \frac{180}{\pi} = \frac{180\pi}{4\pi} = 45$$

Trigonometric Functions

☆ Trigonometric functions refer to the relation between the sides and angles of a right triangle. There are 6 trigonometric functions:

☆ Sine (sin), Cosine (cos), Tangent (tan), Secant (sec), Cosecant (csc), and Cotangent (cot)

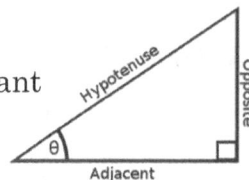

☆ The three main trigonometric functions:

$$SOH - CAH - TOA, \sin\theta = \frac{opposite}{hypotenuse}, Cos\,\theta = \frac{adjacent}{hypotenuse}, tan\,\theta = \frac{opposite}{adjacent}$$

☆ The reciprocal trigonometric functions:

$$csc\,x = \frac{1}{sin\,x}, sec\,x = \frac{1}{cos\,x}, cot\,\theta = \frac{1}{tan\,x}$$

☆ Learn common trigonometric functions:

θ	0°	30°	45°	60°	90°
$\sin\theta$	0	$\frac{1}{2}$	$\frac{\sqrt{2}}{2}$	$\frac{\sqrt{3}}{2}$	1
$\cos\theta$	1	$\frac{\sqrt{3}}{2}$	$\frac{\sqrt{2}}{2}$	$\frac{1}{2}$	0
$\tan\theta$	0	$\frac{\sqrt{3}}{3}$	1	$\sqrt{3}$	Undefined

Examples:

Find each trigonometric function.

Example 1. $sin\,120°$

Solution: Use the following property: $sin(x) = cos(90° - x)$

$sin\,120° = cos(90° - 120°) = cos(-30°) = \frac{\sqrt{3}}{2}$

Example 2. $sin\,135°$

Solution: Use the following property: $sin(x) = cos(90° - x)$

$sin\,135° = cos(90° - 135°) = cos(-45°) = \frac{\sqrt{2}}{2}$

Coterminal Angles and Reference Angles

☆ Coterminal angles are equal angles.

☆ To find a Coterminal of an angle, add or subtract 360 degrees (or 2π for radians) to the given angle.

☆ Reference angle is the smallest angle that you can make from the terminal side of an angle with the x-axis.

Examples:

Example 1. Find a positive and a negative Coterminal angle to angle 65°.

Solution:

$65° - 360° = -295°$
$65° + 360° = 425°$
$-295°$ and a 425° are Coterminal with angle 65°.

Example 2. Find positive and negative Coterminal angles to angle $\frac{\pi}{2}$.

Solution: $\frac{\pi}{2} + 2\pi = \frac{5\pi}{2}$, $\frac{\pi}{2} - 2\pi = -\frac{3\pi}{2}$

Example 3. Find a positive and a negative Coterminal angle to angle 55°.

Solution:

$55° - 360° = -305°$
$55° + 360° = 415°$
$-305°$ and a 415° are Coterminal with angle 55°.

Example 4. Find positive and negative Coterminal angles to angle $\frac{\pi}{3}$.

Solution:

$\frac{\pi}{3} + 2\pi = \frac{7\pi}{3}$, $\frac{\pi}{3} - 2\pi = -\frac{5\pi}{3}$

Evaluating Trigonometric Functions

☆ **Step 1:** Find the reference angle. (It is the smallest angle that you can make from the terminal side of an angle with the x-axis.)

☆ **Step 2:** Determine the quadrant of the function. Depending on the quadrant in which the function lies, the answer will be either positive or negative.

☆ **Step 3:** Find the trigonometric function of the reference angle.

Examples:

Example 1. Find the exact value of trigonometric function. $tan \frac{5\pi}{4}$

Solution: Rewrite the angle for $\frac{5\pi}{4}$:

$tan \frac{5\pi}{4} = tan \left(\frac{4\pi+\pi}{4}\right) = tan \left(\pi + \frac{1}{4}\pi\right)$

Use the periodicity of tan: $tan(x + \pi . k) = tan(x)$

$tan \left(\pi + \frac{1}{4}\pi\right) = tan \left(\frac{1}{4}\pi\right) = 1$

Example 2. Find the exact value of trigonometric function. $sin \frac{7\pi}{6}$

Solution: Rewrite the $sin \frac{7\pi}{6}$.

$sin \frac{7\pi}{6} = sin \left(\frac{\pi}{6} + \pi\right)$

Use the periodicity of sin: $sin(x + \pi . k) = -sin(x)$

$sin \left(\frac{\pi}{6} + \pi\right) = - sin \left(\frac{\pi}{6}\right) = -\frac{1}{2}$

Example 3. Find the exact value of trigonometric function. $tan \frac{4\pi}{3}$

Solution: Rewrite the angle for $\frac{4\pi}{3}$:

$tan \frac{4\pi}{3} = tan \left(\frac{3\pi+\pi}{3}\right) = tan \left(\pi + \frac{1}{3}\pi\right)$

Use the periodicity of tan: $tan(x + \pi . k) = tan(x)$

$tan \left(\pi + \frac{1}{3}\pi\right) = tan \left(\frac{1}{3}\pi\right) = \sqrt{3}$

Missing Sides and Angles of a Right Triangle

☆ By using three main trigonometric functions (Sine, Cosine or Tangent), we can find an unknown side in a right triangle when we have one length, and one angle (apart from the right angle).

☆ A right triangle with Adjacent and Opposite sides and Hypotenuse is shown below.

☆ Recall the three main trigonometric functions:

SOH – CAH – TOA, $sin\,\theta = \frac{opposite}{hypotenuse}$, $Cos\,\theta = \frac{adjacent}{hypotenuse}$, $tan\,\theta = \frac{opposite}{adjacent}$

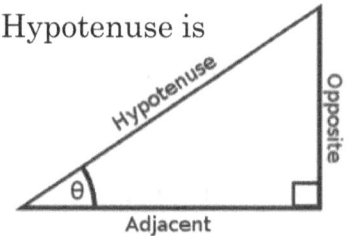

☆ To find missing angles, use inverse of trigonometric functions (examples: $sin^{-1}, cos^{-1},$ and tan^{-1})

Examples:

Example 1. Find side AC in the following triangle. Round your answer to the nearest tenth.

Solution: $sin\,\theta = \frac{opposite}{hypotenuse}$. $sine\,50° = \frac{AC}{6} \rightarrow 6 \times sin\,50° = AC$,

Now use a calculator to find $sine\,50°$.

$$sin\,50° \approx 0.766$$

$AC = 6 \times 0.766 = 4.596$, rounding to the nearest tenth: $4.596 \approx 4.6$

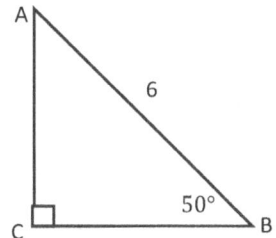

Example 2. Find the value of x in the following triangle.

Solution: $cos\,\theta = \frac{adjacent}{hypotenuse} \rightarrow cos\,x = \frac{10}{14} = \frac{5}{7}$

Use a calculator to find inverse cosine:

$cos^{-1}\left(\frac{5}{7}\right) = 44.41° \approx 44°$, Then: $x = 44$

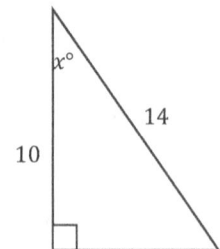

Day 10: Practices

🖎 *Simplify each expression.*

1) $\dfrac{\frac{2}{5}}{\frac{4}{7}} =$

3) $\dfrac{1-\frac{2}{x-1}}{1+\frac{4}{x+1}} =$

2) $\dfrac{6}{\frac{5}{x}+\frac{2}{3x}} =$

4) $\dfrac{x}{\frac{3}{4}-\frac{5}{x}} =$

🖎 *Graph rational function.*

5) $f(x) = \dfrac{x^2-x+2}{x-1}$

🖎 *Simplify each expression.*

6) $\dfrac{5}{x+2} + \dfrac{x-1}{x+2} =$

8) $\dfrac{7}{4x+10} + \dfrac{x-5}{4x+10} =$

7) $\dfrac{6}{x+5} - \dfrac{5}{x+5} =$

🖎 Simplify each expression.

9) $\dfrac{x+1}{x+5} \times \dfrac{x+6}{x+1} =$

12) $\dfrac{x+5}{x+1} \times \dfrac{x^2}{x+5} =$

10) $\dfrac{x+4}{x+9} \times \dfrac{x+9}{x+3} =$

13) $\dfrac{x-3}{x+2} \times \dfrac{2x+4}{x+4} =$

11) $\dfrac{x+8}{x} \times \dfrac{2}{x+8} =$

14) $\dfrac{x-6}{x+3} \times \dfrac{2x+6}{2x} =$

✍ Solve.

15) $\frac{5x}{4} \div \frac{5}{2} =$

16) $\frac{8}{3x} \div \frac{24}{x} =$

17) $\frac{3x}{x+4} \div \frac{x}{3x+12} =$

18) $\frac{2}{5x} \div \frac{16}{10x} =$

19) $\frac{36x}{5} \div \frac{4}{3} =$

20) $\frac{15x^2}{6} \div \frac{5x}{14} =$

✍ Solve each equation.

21) $\frac{1}{8x^2} = \frac{1}{4x^2} - \frac{1}{x} \rightarrow x =$ _____

22) $\frac{1}{x} + \frac{1}{9x} = \frac{5}{36} \rightarrow x =$ _____

23) $\frac{32}{2x^2} + 1 = \frac{8}{x} \rightarrow x =$ _____

24) $\frac{1}{x-5} = \frac{4}{x-5} + 1 \rightarrow x =$ _____

✍ Convert each degree measure into radians.

25) $135° =$

26) $80° =$

27) $270° =$

28) $92° =$

✍ Evaluate.

29) $\sin 90° =$ _____

30) $\sin -330° =$ _____

31) $\cot \frac{2\pi}{3} =$ _____

32) $\tan \frac{\pi}{3} =$ _____

✍ Find a positive and a negative Coterminal angle for each angle.

33) $140° =$
Positive = _____
Negative = _____

34) $-165° =$
Positive = _____
Negative = _____

35) $\frac{5\pi}{4} =$

Positive = _____
Negative = _____

36) $-\frac{7\pi}{9} =$
Positive = _____
Negative = _____

Find the exact value of each trigonometric function.

37) $\cos 180° =$ _____

38) $\cos -270° =$ _____

39) $\tan 225° =$ _____

40) $\sin \frac{\pi}{4} =$ _____

41) $\csc 330° =$ _____

42) $\tan -120° =$ _____

Find the value of x in each triangle.

43) _____

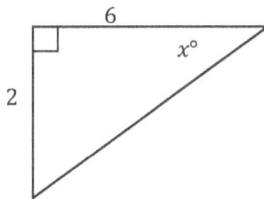

44) _____

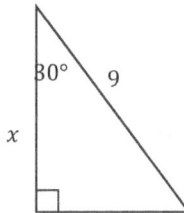

45) _____

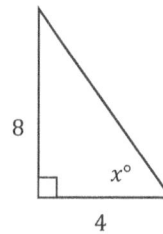

Day 10: Answers

1) $\frac{7}{10}$ (First, write the complex fraction using the division sign: $\frac{\frac{2}{5}}{\frac{4}{7}} = \frac{2}{5} \div \frac{4}{7}$. Use the dividing fractions rule: Keep, Change, Flip (keep the first fraction, change the division sign to multiplication, flip the second fraction) $\frac{2}{5} \div \frac{4}{7} = \frac{2}{5} \times \frac{7}{4} = \frac{14}{20} = \frac{7}{10}$)

2) $\frac{18x}{17}$ (First, simplify the denominator: $\frac{5}{x} + \frac{2}{3x} = \frac{17}{3x}$, then, write the complex fraction using the division sign: $\frac{6}{\frac{5}{x} + \frac{2}{3x}} = \frac{6}{\frac{17}{3x}} = 6 \div \frac{17}{3x}$. Use the dividing fractions rule: Keep, Change, Flip (keep the first fraction, change the division sign to multiplication, flip the second fraction) $6 \div \frac{17}{3x} = 6 \times \frac{3x}{17} = \frac{18x}{17}$)

3) $\frac{x^2 - 2x - 3}{x^2 + 4x - 5}$ (First, simplify the numerator: $1 - \frac{2}{x-1} = \frac{x-3}{x-1}$, then, simplify the denominator: $1 + \frac{4}{x+1} = \frac{x+5}{x+1}$, Now, write the complex fraction using the division sign ($\div$): $\frac{1 - \frac{2}{x-1}}{1 + \frac{4}{x+1}} = \frac{\frac{x-3}{x-1}}{\frac{x+5}{x+1}} = \frac{x-3}{x-1} \div \frac{x+5}{x+1}$. Use the dividing fractions rule: Keep, Change, Flip (keep the first fraction, change the division sign to multiplication, flip the second fraction) $\frac{x-3}{x-1} \div \frac{x+5}{x+1} = \frac{x-3}{x-1} \times \frac{x+1}{x+5} = \frac{(x-3)(x+1)}{(x-1)(x+5)} = \frac{x^2 - 2x - 3}{x^2 + 4x - 5}$)

4) $\frac{4x^2}{3x - 20}$ (First, simplify the denominator: $\frac{3}{4} - \frac{5}{x} = \frac{3x - 20}{4x}$, then, write the complex fraction using the division sign: $\frac{x}{\frac{3}{4} - \frac{5}{x}} = \frac{x}{\frac{3x-20}{4x}} = x \div \frac{3x-20}{4x}$. Use the dividing fractions rule: Keep, Change, Flip (keep the first fraction, change the division sign to multiplication, flip the second fraction) $x \div \frac{3x-20}{4x} = x \times \frac{4x}{3x-20} = \frac{4x^2}{3x-20}$)

5) First, notice that the graph is in two pieces. Most rational functions have graphs in multiple pieces. Find $y - $ intercept by substituting zero for x and solving for y ($f(x)$): $x = 0 \rightarrow y = \frac{x^2 - x + 2}{x - 1} = \frac{0^2 - 0 + 2}{0 - 1} = -2$,

$y - $ intercept: $(0, -2)$ Asymptotes of $\frac{x^2 - x + 2}{x - 1}$: Vertical: $x = 1$, Slant asymptote: $y = 2x + 1$ (divide the numerator by the denominator). After finding the asymptotes, you can plug in some values for x and solve for y. Here is the sketch for this function.

6) $\frac{x+4}{x+2}$ (The denominators are equal. Then, use fractions addition rule: $\frac{a}{c} \pm \frac{b}{c} = \frac{a \pm b}{c} \rightarrow \frac{5}{x+2} + \frac{x-1}{x+2} = \frac{5+(x-1)}{x+2} = \frac{x+4}{x+2}$)

7) $\frac{1}{x+5}$ (The denominators are equal. Then, use fractions addition rule: $\frac{a}{c} \pm \frac{b}{c} = \frac{a \pm b}{c} \rightarrow \frac{6}{x+5} - \frac{5}{x+5} = \frac{6-5}{x+5} = \frac{1}{x+5}$)

8) $\frac{x+2}{4x+10}$ (The denominators are equal. Then, use fractions addition rule: $\frac{a}{c} \pm \frac{b}{c} = \frac{a \pm b}{c} \rightarrow \frac{7}{4x+10} + \frac{x-5}{4x+10} = \frac{7+(x-5)}{4x+10} = \frac{x+2}{4x+10}$)

9) $\frac{x+6}{x+5}$ (Multiply numerators and denominators: $\frac{a}{b} \times \frac{c}{d} = \frac{a \times c}{b \times d} \rightarrow \frac{x+1}{x+5} \times \frac{x+6}{x+1} = \frac{(x+1)(x+6)}{(x+5)(x+1)}$. Cancel the common factor: $(x+1)$. Then: $\frac{(x+1)(x+6)}{(x+5)(x+1)} = \frac{x+6}{x+5}$)

10) $\frac{x+4}{x+3}$ (Multiply numerators and denominators: $\frac{a}{b} \times \frac{c}{d} = \frac{a \times c}{b \times d} \rightarrow \frac{x+4}{x+9} \times \frac{x+9}{x+3} = \frac{(x+4)(x+9)}{(x+9)(x+3)}$. Cancel the common factor: $(x+9)$. Then: $\frac{(x+4)(x+9)}{(x+9)(x+3)} = \frac{x+4}{x+3}$)

11) $\frac{2}{x}$ (Multiply numerators and denominators: $\frac{a}{b} \times \frac{c}{d} = \frac{a \times c}{b \times d} \rightarrow \frac{x+8}{x} \times \frac{2}{x+8} = \frac{(x+8)(2)}{(x)(x+8)}$. Cancel the common factor: $(x+8)$. Then: $\frac{(x+8)(2)}{(x)(x+8)} = \frac{2}{x}$)

12) $\frac{x^2}{x+1}$ (Multiply numerators and denominators: $\frac{a}{b} \times \frac{c}{d} = \frac{a \times c}{b \times d} \rightarrow \frac{x+5}{x+1} \times \frac{x^2}{x+5} = \frac{(x+5)(x^2)}{(x+1)(x+5)}$. Cancel the common factor: $(x+5)$. Then: $\frac{(x+5)(x^2)}{(x+1)(x+5)} = \frac{x^2}{x+1}$)

13) $\frac{2(x-3)}{x+4}$ (Multiply numerators and denominators: $\frac{a}{b} \times \frac{c}{d} = \frac{a \times c}{b \times d} \rightarrow \frac{x-3}{x+2} \times \frac{2x+4}{x+4} = \frac{(x-3)(2x+4)}{(x+2)(x+4)}$. Factor $2x+4 = 2(x+2)$ Cancel the common factor: $(x+2)$. Then: $\frac{(x-3)(2)(x+2)}{(x+2)(x+4)} = \frac{2(x-3)}{x+4}$)

14) $\frac{x-6}{x}$ (Multiply numerators and denominators: $\frac{a}{b} \times \frac{c}{d} = \frac{a \times c}{b \times d} \rightarrow \frac{x-6}{x+3} \times \frac{2x+6}{2x} = \frac{(x-6)(2x+6)}{(x+3)(2x)}$. Factor $2x+6 = 2(x+3)$ Cancel the common factor: $2(x+3)$. Then: $\frac{(x-6)(2)(x+3)}{(x+3)(2x)} = \frac{x-6}{x}$)

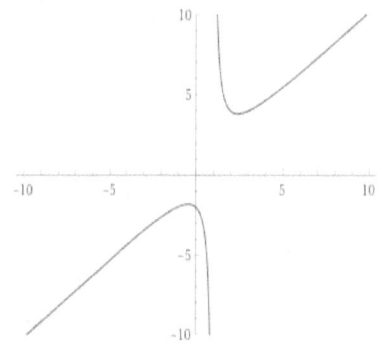

15) $\frac{x}{2}$ (Use fractions division rule: $\frac{a}{b} \div \frac{c}{d} = \frac{a}{b} \times \frac{d}{c} = \frac{a \times d}{b \times c} \rightarrow \frac{5x}{4} \div \frac{5}{2} = \frac{5x}{4} \times \frac{2}{5} = \frac{2(5x)}{4 \times 5}$. Now, cancel common factors. Then: $\frac{x}{2}$)

16) $\frac{1}{9}$ (Use fractions division rule: $\frac{a}{b} \div \frac{c}{d} = \frac{a}{b} \times \frac{d}{c} = \frac{a \times d}{b \times c} \rightarrow \frac{8}{3x} \div \frac{24}{x} = \frac{8}{3x} \times \frac{x}{24} = \frac{8x}{(3x)(24)}$. Now, cancel common factors. Then: $\frac{8x}{(3x)(24)} = \frac{1}{9}$)

17) 9 (Use fractions division rule: $\frac{a}{b} \div \frac{c}{d} = \frac{a}{b} \times \frac{d}{c} = \frac{a \times d}{b \times c} \rightarrow \frac{3x}{x+4} \div \frac{x}{3x+12} = \frac{3x}{x+4} \times \frac{3x+12}{x} = \frac{(3x)(3x+12)}{(x+4)(x)}$. Now, factorize the expressions $3x + 12$. Then: $3x + 12 = 3(x + 4)$. Simplify: $\frac{(3x)(3x+12)}{(x+4)(x)} = \frac{(3x)(3)(x+4)}{(x+4)(x)}$, cancel common factors. Then: $\frac{(3x)(3)(x+4)}{(x+4)(x)} = 9$)

18) $\frac{1}{4}$ (Use fractions division rule: $\frac{a}{b} \div \frac{c}{d} = \frac{a}{b} \times \frac{d}{c} = \frac{a \times d}{b \times c} \rightarrow \frac{2}{5x} \div \frac{16}{10x} = \frac{2}{5x} \times \frac{10x}{16} = \frac{2(10x)}{(5x)(16)}$. Now, cancel common factors. Then: $\frac{2(10x)}{(5x)(16)} = \frac{1}{4}$)

19) $\frac{27x}{5}$ (Use fractions division rule: $\frac{a}{b} \div \frac{c}{d} = \frac{a}{b} \times \frac{d}{c} = \frac{a \times d}{b \times c} \rightarrow \frac{36x}{5} \div \frac{4}{3} = \frac{36x}{5} \times \frac{3}{4} = \frac{3(36x)}{5 \times 4}$. Now, cancel common factors. Then: $\frac{3(36x)}{5 \times 4} = \frac{27x}{5}$)

20) $7x$ (Use fractions division rule: $\frac{a}{b} \div \frac{c}{d} = \frac{a}{b} \times \frac{d}{c} = \frac{a \times d}{b \times c} \rightarrow \frac{15x^2}{6} \div \frac{5x}{14} = \frac{15x^2}{6} \times \frac{14}{5x} = \frac{14(15x^2)}{6(5x)}$. Now, cancel common factors. Then: $\frac{14(15x^2)}{6(5x)} = 7x$)

21) $x = \frac{1}{8}$ (First, simplify right side of the equation: $\frac{1}{4x^2} - \frac{1}{x} = \frac{1-4x}{4x^2}$. Use cross multiply method: if $\frac{a}{b} = \frac{c}{d}$, then: $a \times d = b \times c \rightarrow \frac{1}{8x^2} = \frac{1-4x}{4x^2} \rightarrow 4x^2 = (8x^2)(1 - 4x)$. Expand: $(8x^2)(1 - 4x) = 8x^2 - 32x^3$, Then: $4x^2 = 8x^2 - 32x^3$, Now, subtract both sides $(4x^2)$, Then: $4x^2 - (4x^2) = 8x^2 - 32x^3 - (4x^2) \rightarrow 0 = 4x^2 - 32x^3 \rightarrow x^2(4 - 32x) = 0 \rightarrow x = \frac{1}{8}$)

22) $x = 8$ (First, simplify left side of the equation: $\frac{1}{x} + \frac{1}{9x} = \frac{10}{9x}$. Use cross multiply method: if $\frac{a}{b} = \frac{c}{d}$, then: $a \times d = b \times c \rightarrow \frac{10}{9x} = \frac{5}{36} \rightarrow 10 \times 36 = 5(9x)$. Then: $360 = 45x \rightarrow x = \frac{360}{45} = 8$)

23) $x = 4$ (First, simplify left side of the equation: $\frac{32}{2x^2} + 1 = \frac{32+2x^2}{2x^2}$. Use cross multiply method: if $\frac{a}{b} = \frac{c}{d}$, then: $a \times d = b \times c \rightarrow \frac{32+2x^2}{2x^2} = \frac{8}{x} \rightarrow (32 + 2x^2)(x) = 8(2x^2)$. Expand: $(32 + 2x^2)(x) = 32x + 2x^3$, Then: $32x + 2x^3 = 16x^2$, Now, subtract both sides $(16x^2)$, Then: $32x + 2x^3 - 16x^2 = 16x^2 - 16x^2 \rightarrow 2x^3 - 16x^2 + 32x = 0 \rightarrow 2x(x^2 - 8x + 16) = 0 \rightarrow 2x(x - 4)^2 \rightarrow x = 4$)

24) $x = 2$ (First, simplify right side of the equation: $\frac{4}{x-5} + 1 = \frac{x-1}{x-5}$. Use cross multiply method: if $\frac{a}{b} = \frac{c}{d}$, then: $a \times d = b \times c \rightarrow \frac{1}{x-5} = \frac{x-1}{x-5}$. the denominators on both side of the equation are equal. Therefore, their numerators must be equal too. $x - 1 = 1 \rightarrow x = 2$)

25) $\frac{3\pi}{4}$ (Use this formula: Radians = Degrees $\times \frac{\pi}{180}$. Radians = $135 \times \frac{\pi}{180} = \frac{135\pi}{180} = \frac{3\pi}{4}$)

26) $\frac{4\pi}{9}$ (Use this formula: Radians = Degrees $\times \frac{\pi}{180}$. Radians = $80 \times \frac{\pi}{180} = \frac{80\pi}{180} = \frac{4\pi}{9}$)

27) $\frac{4\pi}{9}$ (Use this formula: Radians = Degrees $\times \frac{\pi}{180}$. Radians = $270 \times \frac{\pi}{180} = \frac{270\pi}{180} = \frac{3\pi}{2}$)

28) $\frac{23\pi}{45}$ (Use this formula: Radians = Degrees $\times \frac{\pi}{180}$. Radians = $92 \times \frac{\pi}{180} = \frac{92\pi}{180} = \frac{23\pi}{45}$)

29) 1 (According to the rules of trigonometry $sin\ 90° = 1$)

30) $\frac{1}{2}$ (Use the following property: $sin(x) = sin(-360° + x) \rightarrow sin(-330°) = sin(-360° + 30°) = sin(30°) = \frac{1}{2}$)

31) $-\frac{\sqrt{3}}{3}$ (Use the following property: $cot(x) = \frac{cos(x)}{sin(x)} \rightarrow cot\left(\frac{2\pi}{3}\right) = \frac{cos\left(\frac{2\pi}{3}\right)}{sin\left(\frac{2\pi}{3}\right)} = \frac{cos\left(\pi - \frac{\pi}{3}\right)}{sin\left(\pi - \frac{\pi}{3}\right)} = \frac{-\frac{1}{2}}{\frac{\sqrt{3}}{2}} = -\frac{\sqrt{3}}{3}$)

32) $\sqrt{3}$ (Use the following property: $tan(x) = \frac{sin(x)}{cos(x)} = tan\left(\frac{\pi}{3}\right) = \frac{sin\left(\frac{\pi}{3}\right)}{cos\left(\frac{\pi}{3}\right)} = \frac{\frac{\sqrt{3}}{2}}{\frac{1}{2}} = \sqrt{3}$)

33) Positive = 500° , Negative = −220° ($140° − 360° = −220°$
   $65° + 360° = 500°$
   −220° and a 500° are Coterminal with angle 140°)

34) Positive = 195° , Negative = −525° ($−165° − 360° = −525°$
   $−165° + 360° = 195°$
   −525° and a 195° are Coterminal with angle −165°)

35) Positive = $\frac{13\pi}{4}$, Negative = $-\frac{3\pi}{4}$ ($\frac{5\pi}{4} − 2\pi = −\frac{3\pi}{4}$
   $\frac{5\pi}{4} + 2\pi = \frac{13\pi}{4}$
   $-\frac{3\pi}{4}$ and a $\frac{13\pi}{4}$ are Coterminal with angle $\frac{5\pi}{4}$)

36) Positive $=\frac{11\pi}{9}$, Negative $=-\frac{25\pi}{9}$ $(-\frac{7\pi}{9} - 2\pi = -\frac{25\pi}{9}$

$-\frac{7\pi}{9} + 2\pi = \frac{11\pi}{9}$

$-\frac{25\pi}{9}$ and a $\frac{11\pi}{9}$ are Coterminal with angle $-\frac{7\pi}{9})$

37) -1 (According to the rules of trigonometry $cos\,(180° + \theta) = -cos\,(\theta) \rightarrow$ $cos\,(180° + 0) = -cos\,(0) = -1)$

38) 0 (According to the rules of trigonometry $cos\,(-270°) = cos\,(90°) = 0)$

39) 1 (Rewrite the angle for 225 : $tan\,225° = tan\,(180° + 45°) =$. Use the periodicity of tan: $tan(x + \pi.k) = tan(x) \rightarrow tan(180° + 45°) = tan(45°) = 1)$

40) $\frac{\sqrt{2}}{2}$ (According to the rules of trigonometry $sin\frac{\pi}{4} = \frac{\sqrt{2}}{2})$

41) -2 (Use the following property:

$csc\,x = \frac{1}{sin\,x} \rightarrow csc\,330° = \frac{1}{sin\,330°} = \frac{1}{-sin\,30°} = \frac{1}{-\frac{1}{2}} = -2)$

42) $\sqrt{3}$ (Rewrite the angle for 120 : $tan - 120° = tan\,(-180° + 60°) =$. Use the periodicity of tan: $tan(x + \pi.k) = tan(x) \rightarrow tan(-180° + 60°) = tan(60°) = \sqrt{3})$

43) 18 ($tan\,\theta = \frac{opposite}{adjacent}$. $tan\,x° = \frac{2}{6} \rightarrow x = 18$)

44) 7.8 ($Cos\,\theta = \frac{adjacent}{hypotenuse}$. $Cos\,30° = \frac{x}{9} \rightarrow x = 9 \times Cos\,30°$, we know that $Cos\,30° = \frac{\sqrt{3}}{2} \approx 0.87 \rightarrow x = 9 \times \frac{\sqrt{3}}{2} = 9 \times 0.87 = 7.83$, rounding to the nearest tenth: $7.83 \approx 7.8$)

45) 63 ($tan\,\theta = \frac{opposite}{adjacent}$. $tan\,x° = \frac{8}{4} \rightarrow x = 63$)

Time to Test

Time to refine your skill with a practice examination

In this section, there are 5 complete ALEKS Mathematics Tests. Take these tests to simulate the test day experience. After you've finished, score your test using the answers and explanations section.

Before You Start

- You'll need a pencil and scratch papers to take the test.
- For these practice tests, don't time yourself. Spend time as much as you need.
- After you've finished the test, review the answer key to see where you went wrong.

Good luck!

ALEKS Mathematics

Practice Test 1

2023 – 2024

Total number of questions: 35

Total time: No time limit

Calculators are permitted for ALEKS Math Test.

1) Factor.

$4x^2 - 49$

2) Solve for x.

$7x^2 - 28x = 0$

3) Two consecutive odd integers have a sum of 28. Find the integers.

4) The equation of a line is given below. $-8x - 2y = 4$
Find the slope and y-intercept.
Then use them to graph the line.
Slope: _____

y-intercept: _____

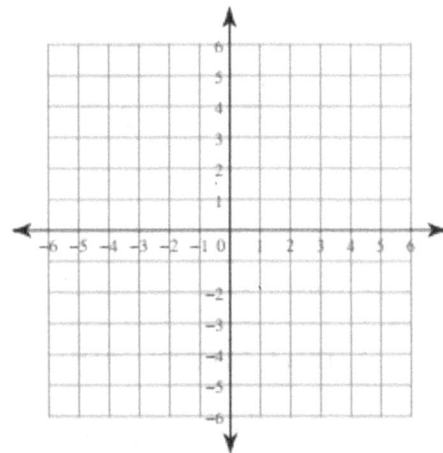

5) Find the domain and range for $f(x) = x^2 + 2$

6) What is the value of y in the following system of equations?

$$2x + 5y = 11$$
$$4x - 2y = -14$$

7) Solve the equation: $log_3(x + 20) - log_3(x + 2) = 1$

8) Write as a single fraction.

$$-4 + \frac{2a - 6y}{12a} - \frac{6a + 4y}{8a}$$

Simplify your answer as much as possible.

9) If $f(x) = 2x^3 + 5x^2 + 2x$ and $g(x) = -3$, what is the value of $f(g(x))$?

10) If the area of a circle is 64 square meters, what is its radius?

11) If one angle of a right triangle measures 60°, what is the sine of the other acute angle?

12) In the figure below, line A is parallel to line B. What is the value of angle x?

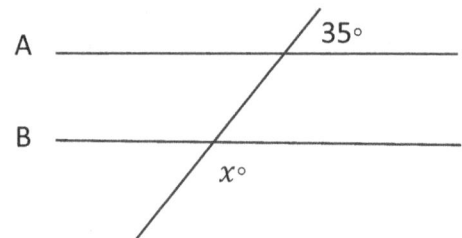

A ———————————— 35°

B ————————————

$x°$

13) An angle is equal to one fifth of its supplement. What is the measure of that angle?

14) Simplify $\frac{4-3i}{-4i}$?

15) The average of five consecutive numbers is 40. What is the smallest number?

16) If $sin A = \frac{1}{4}$ in a right triangle and the angle A is an acute angle, then what is $cos A$?

17) In the standard (x, y) coordinate system plane, what is the area of the circle with the following equation?

$$(x + 2)^2 + (y - 4)^2 = 16$$

18) What is the slope of a line that is perpendicular to the line

$$4x - 2y = 14?$$

19) If $f(x) = 2x^3 + 4$ and $g(x) = \frac{1}{x}$, what is the value of $f(g(x))$?

20) If 150% of a number is 75, then what is 90% of that number?

21) If cotangent of an angel β is 1, then the tangent of angle β is ...

22) Simplify this expression: $\sqrt{\frac{x^2}{2} + \frac{x^2}{16}}$?

23) If $\tan x = \frac{8}{15}$, then $\sin x =$

$$f(x) = \frac{1}{(x-3)^2 + 4(x-3) + 4}$$

24) For what value of x is the function $f(x)$ above undefinded?

25) What are the zeroes of the function $f(x) = x^3 + 6x^2 + 8x$?

26) Simplify as much as possible: $\frac{5}{x^2} + \frac{7x-3}{x^3}$

27) In the following figure, point Q lies on line n, what is the value of y if $x = 35$?

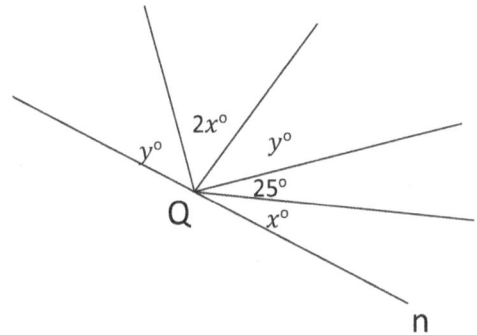

28) Find the answer. $(3n^2 + 4n + 6) - (2n^2 - 5)$

29) The diagonal of a rectangle is 10 inches long and the height of the rectangle is 8 inches. What is the perimeter of the rectangle?

30) Sara opened a bank account that earns 2 percent compounded annually. Her initial deposit was \$1,500, and she uses the expression $\$1,500(x)^n$ to find the value of the account after n years. What is the value of x in the expression?

31) Sketch the graph of $y + 3 = 2(x - 1)$.

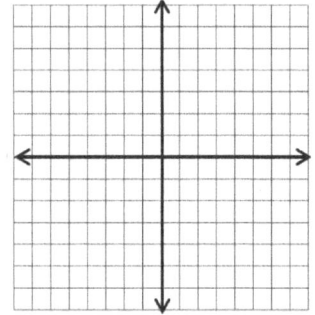

32) Perform the indicated operation and write the result in standard form: $(-7 + 12i)(-5 - 15i)$

33) Write in terms of $log(r), log(s), log(t)$: $log\,\frac{s\sqrt{t}}{r^2}$

34) What is the least common denominator for $\frac{3x}{x^2-36}$ and $\frac{4}{2x-12}$?

35) Find AC in the following triangle. Round your answer to the nearest tenth.

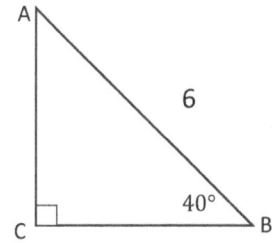

This is the end of Practice Test 1.

ALEKS Mathematics

Practice Test 2

2023 – 2024

Total number of questions: 35

Total time: No time limit

Calculators are permitted for ALEKS Math Test.

1) Simplify $(-3 + 9i)(3 + 5i)$.

2) Multiply.
$3x^4u^5 \times 5u^2 \times 2x$
Simplify your answer as much as possible.

3) Solve for u.
$3u^2 + 15u = 0$

4) Factor by grouping (sometime called the ac-method).
$4x^2 - 4x - 15$
First, choose a form with appropriate signs.
Then, fill in the blanks with numbers to be used for grouping.
Finally, show the factorization.

5) Find the domain and the rage of function: $f(x) = \frac{1}{x+2}$

6) $(x - 5)(x^2 + 5x + 4) = ?$

7) $5 + 8 \times (-3) - [4 + 22 \times 5] \div 6 = ?$

8) Simplify. $\dfrac{\dfrac{1}{2} - \dfrac{x+5}{4}}{\dfrac{x^2}{2} - \dfrac{5}{2}}$

9) Find $tan \dfrac{2\pi}{3}$

10) What is the value of x in this equation?

$$4\sqrt{2x + 6} = 24$$

11) What is the center and radius of a circle with the following equation?

$$(x - 4)^2 + (y + 7)^2 = 3$$

12) If the center of a circle is at the point $(-4, 2)$ and its circumference equals to 2π, what is the standard form equation of the circle?

13) Find the value of x in the following diagram.

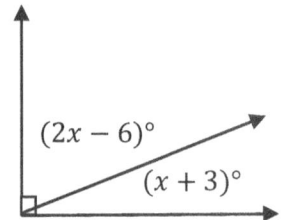

14) If $\frac{1}{m+n} = \frac{1}{t^2}$, then $n =$

15) Find a positive and a negative Coterminal angle to angle 125°.

Positive coterminal angle of 125°: _____

Negative coterminal angle of 125°: _____

16) Solve inequality and graph it. $8x - 4 \geq -2x + 16$

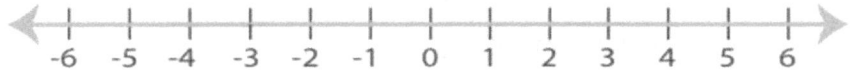

17) Find the factors of $x^2 - 7x + 12$.

18) Solve for x: $2(12x + 6) = -2(x + 1)$

19) What is the distance between the points $(1, 3)$ and $(-2, 7)$ on the coordinate plane?

20) The interval solution to the inequality $\frac{6x+12}{x-8} > 0$ is:

Solution: _____

Interval Notation: _____

21) The average of five numbers is 26. If a sixth number 56 is added, then, what is the new average?

22) Simplify the expression ($x \neq -4$ and $x \neq 6$): $\dfrac{1}{\frac{1}{x-6}+\frac{1}{x+4}}$

23) What is the equation (in standard form) of the following graph?

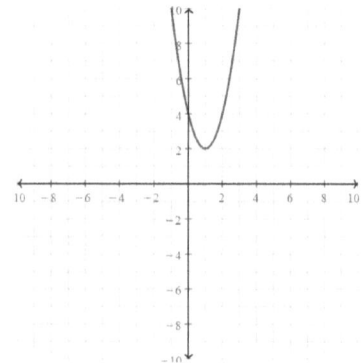

24) Simplify: $\dfrac{4x^3 + 2x^2 y - 4xy^2}{2x^2 - 2xy}$

25) Find the solutions of the following equation.

$$x^2 + 2x - 5 = 0$$

26) *Sketch the graph of linear inequality:* $y \leq -2x - 2$

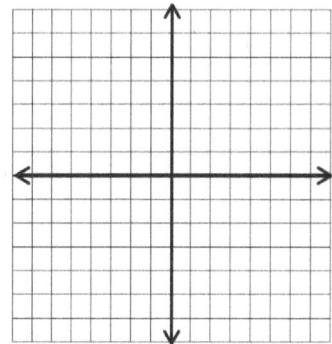

27) The profit (in dollars) from a carwash is given by the function $P(a)$,
$P(a) = \dfrac{40a - 500}{a} + b$, where a is the number of cars washed and b is a constant?
If 50 cars were washed today for a total profit of \$600, what is the value of b?

28) If $log_2 x = 5$, then $x = $?

29) What is the value of $\frac{8b}{c}$ when $\frac{c}{b} = 2$?

30) What is the equivalent temperature of $140°F$ in Celsius?

$$C = \frac{5}{9}(F - 32)$$

31) $\frac{2x}{x+8} - \frac{7}{x-8} = $

32) Find the equation of the horizontal asymptote of the function $f(x) = \frac{x+3}{x^2+1}$.

33) Write Find the value of x in the following triangle. (round your answer to the whole number)

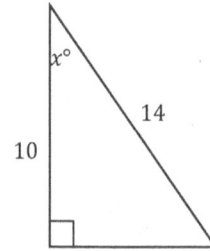

34) Convert 150 degrees to radian.

35) Solve and write the answer in scientific notation: $(6.4 \times 10^7) - (3.2 \times 10^6)$

This is the end of Practice Test 2.

ALEKS Mathematics
Practice Tests Answers
and Explanations

ALEKS Mathematics Practice Test 1
Answers and Explanations

1) **The answer is** $(2x - 7)(2x + 7)$

Rewrite 4 as 2^2: $4x^2 = 2^2x^2$. Rewrite 49 as 7^2. Now, apply exponent rule: $a^m b^m = (ab)^m$. Then: $2^2x^2 = (2x)^2$. Apply difference of two square formula: $x^2 - y^2 = (x - y)(x + y) \rightarrow 4x^2 - 49 = (2x)^2 - 7^2 = (2x - 7)(2x + 7)$

2) **The answer is** $x = 0$ **or** $x = 4$

To solve for x, divide both sides by 7. Then: $7x^2 - 28x = 0 \rightarrow \frac{7x^2}{7} - \frac{28x}{7} = \frac{0}{7} \rightarrow$ $x^2 - 4x = 0$. Now, take the common factor x out: $x^2 - 4x = 0 \rightarrow x(x - 4) = 0$ The product of x and $(x - 4)$ is zero. Therefore, x is zero or $(x - 4)$ is zero. Then: $x = 0$ or $x - 4 = 0 \rightarrow x = 4$

3) **The answers are** **13** *and* **15**

Let's put x for smaller integer. Then, the two integers are x and $x + 2$. (the difference of any two consecutive odd (or even) integers is 2). The sum of two integers is 28. Write the equation and solve for x:

$$x + (x + 2) = 28 \rightarrow 2x + 2 = 28 \rightarrow 2x = 26 \rightarrow \frac{2x}{2} = \frac{26}{2} \rightarrow x = 13$$

The smaller integer is 13 and the bigger integer is 15 ($13 + 2 = 15$).

4) **The answers are: Slope is** -4 **and** y**-intercept is** -2

Write the equation in slope intercept form. The slope intercept form of the equation of a line is: $y = mx + b$. Then: $-8x - 2y = 4 \rightarrow -8x - 2y + 8x = 4 + 8x \rightarrow$

$$-2y = 8x + 4 \rightarrow \frac{-2y}{-2} = \frac{8x}{-2} + \frac{4}{-2} \rightarrow y = -4x - 2 \rightarrow$$

The slope intercept form of the line is: $y = -4x - 2$.

Then, the slope is -4 and the $y-$intercept is -2.

Now, you can graph the line.

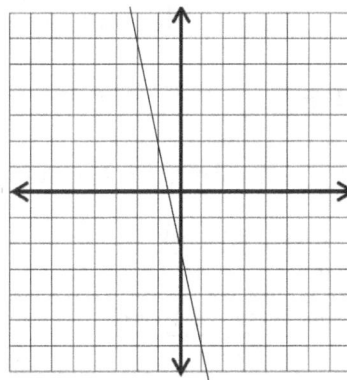

5) The answer is: Domain: all real values of x, range: all real number where $f(x) \geq 2$

The domain of $f(x)$ is "all real values of x. (x can be any real number)

Range: since x^2 is never negative, $x^2 + 2$ is never less than 2.

Hence, the range of $f(x)$ is "all real numbers where $f(x) \geq 2$".

6) The answer is 3

Solving systems of equations by elimination: Multiply the first equation by (-2), then add it to the second equation.

$$\begin{array}{c} -2(2x + 5y = 11) \\ \underline{4x - 2y = -14} \end{array} \Rightarrow \begin{array}{c} -4x - 10y = -22 \\ 4x - 2y = -14 \end{array} \Rightarrow -12y = -36 \Rightarrow y = 3$$

7) The answer is $x = 7$

$log_3(x + 20) - log_3(x + 2) = 1$. First, condense the two logarithms:

$$log_3(x + 20) - log_3(x + 2) = 1 \rightarrow log_3\left(\frac{x + 20}{x + 2}\right) = 1$$

We know that $log_a a = 1$. Then: $log_3\left(\frac{x+20}{x+2}\right) = log_3 3 \rightarrow \frac{x+20}{x+2} = 3$

Now, use cross multiplication and solve for x.

$$\frac{x + 20}{x + 2} = 3 \rightarrow x + 20 = 3(x + 2) \rightarrow x + 20 = 3x + 6 \rightarrow 14 = 2x \rightarrow x = 7$$

8) The answer is $\frac{-12y - 55a}{12a}$

To write this expression as a single fraction, we need to find a common denominator.

The common denominator of $12a$ and $8a$ is $24a$. Then:

$$-4 + \frac{2a-6y}{12a} - \frac{6a+4y}{8a} = \frac{-4(24a)}{24a} + \frac{2(2a-6y)}{24a} - \frac{3(6a+4y)}{24a}$$

Now, simplify the numerators and combine:

$$\frac{-4(24a)}{24a} + \frac{2(2a-6y)}{24a} - \frac{3(6a+4y)}{24a} = \frac{-96a}{24a} + \frac{4a-12y}{24a} - \frac{18a+12y}{24a} =$$

$$\frac{-96a+4a-12y-18a-12y}{24a} = \frac{-110a-24y}{24a}$$

Divide both numerator and denominator by 2. Then:

$$\frac{-110a - 24y}{24a} = \frac{-12y - 55a}{12a}$$

9) The answer is −15

$g(x) = -3$, then $f(g(x)) = f(-3) = 2(-3)^3 + 5(-3)^2 + 2(-3) = -54 + 45 - 6 = -15$

10) The answer is $\frac{8\sqrt{\pi}}{\pi}$

Formula for the area of a circle is: $A = \pi r^2$, using 64 for the area of the circle we have: $64 = \pi r^2$. Solve for the radius (r). $\frac{64}{\pi} = r^2 \rightarrow r = \sqrt{\frac{64}{\pi}} = \frac{8}{\sqrt{\pi}} = \frac{8}{\sqrt{\pi}} \times \frac{\sqrt{\pi}}{\sqrt{\pi}} = \frac{8\sqrt{\pi}}{\pi}$

11) The answer is $\frac{1}{2}$

The relationship among all sides of right triangle $30° - 60° - 90°$ is provided in the following triangle: Sine of $30°$ equals to: $\frac{opposite}{hypotenuse} = \frac{x}{2x} = \frac{1}{2}$

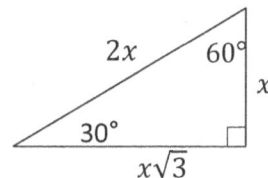

12) The answer is 145°

The angle x and 35 are supplementary angles. Therefore: $x + 35 = 180 \rightarrow$

$$x = 180° - 35° = 145°$$

13) The answer is 30°

The sum of supplement angles is 180. Let x be that angle. Therefore, $x + 5x = 180$

$6x = 180$, divide both sides by 6: $x = 30$ degrees

14) The answer is $\frac{3}{4} + i$

To simplify the fraction, multiply both numerator and denominator by i.

$\frac{4 - 3i}{-4i} \times \frac{i}{i} = \frac{4i - 3i^2}{-4i^2}$, $i^2 = -1$, Then: $\frac{4i - 3i^2}{-4i^2} = \frac{4i - 3(-1)}{-4(-1)} = \frac{4i + 3}{4} = \frac{4i}{4} + \frac{3}{4} = \frac{3}{4} + i$

15) The answer is 38

Let x be the smallest number.

Then, these are the numbers: $x, x + 1, x + 2, x + 3,$

and $x + 4$. $average = \frac{sum\ of\ terms}{number\ of\ terms} \Rightarrow 40 = \frac{x+(x+1)+(x+2)+(x+3)+(x+4)}{5}$

$\Rightarrow 40 = \frac{5x+10}{5} \Rightarrow$

$200 = 5x + 10 \Rightarrow 190 = 5x \Rightarrow x = 38$

The smallest number is 38.

16) The answer is $\frac{\sqrt{15}}{4}$

$sinA = \frac{1}{4} \Rightarrow$ Since $sin\theta = \frac{opposite}{hypotenuse}$, we have the following right triangle. Then:

$c = \sqrt{4^2 - 1^2} = \sqrt{16 - 1} = \sqrt{15}$

$cosA = \frac{adjacent}{hypotenuse} = \frac{\sqrt{15}}{4}$

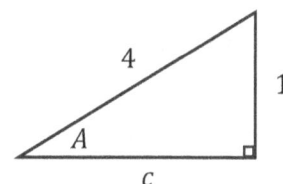

17) The answer is 16π

The equation of a circle in standard form is: $(x - h)^2 + (y - k)^2 = r^2$, where r is the radius of the circle. In this circle the radius is 4. $r^2 = 16 \rightarrow r = 4$, $(x + 2)^2 + (y - 4)^2 = 16$, Area of a circle: $A = \pi r^2 = \pi(4)^2 = 16\pi$

18) The answer is $-\frac{1}{2}$

The equation of a line in slope intercept form is: $y = mx + b$, Solve for y. $4x - 2y = 14 \Rightarrow -2y = 14 - 4x \Rightarrow y = \frac{(14-4x)}{-2} \Rightarrow y = 2x - 7$, The slope of the line is 2. The slope of the line perpendicular to this line is: $m_1 \times m_2 = -1 \Rightarrow 2 \times m_2 = -1 \Rightarrow m_2 = -\frac{1}{2}$

19) The answer is $\frac{2}{x^3} + 4$

$g(x) = \frac{1}{x}$. To find $f(g(x))$, substitute x with $\frac{1}{x}$ in the function $f(x)$. Then:

$$f(g(x)) = f\left(\frac{1}{x}\right) = 2 \times \left(\frac{1}{x}\right)^3 + 4 = \frac{2}{x^3} + 4$$

20) The answer is 45

First, find the number. Let x be the number. Write the equation and solve for x. 150% of a number is 75, then: $1.5 \times x = 75 \Rightarrow x = 75 \div 1.5 = 50$, 90% of 50 is: $0.9 \times 50 = 45$

21) The answer is 1

The cotangent is the reciprocal of tangent: $tangent\ \beta = \frac{1}{cotangent\ \beta} = \frac{1}{1} = 1$

22) The answer is $\frac{3}{4}x$

Find the common denominator and simplify the expression.

$$\sqrt{\frac{x^2}{2} + \frac{x^2}{16}} = \sqrt{\frac{8x^2}{16} + \frac{x^2}{16}} = \sqrt{\frac{9x^2}{16}} = \sqrt{\frac{9}{16}x^2} = \sqrt{\frac{9}{16}} \times \sqrt{x^2} = \frac{3}{4} \times x = \frac{3}{4}x$$

23) The answer is $\frac{8}{17}$

$tan\theta = \frac{opposite}{adjacent}$, and $tan\ x = \frac{8}{15}$, therefore, the opposite side of the angle x is 8 and the adjacent side is 15. Let's draw the triangle.

Using Pythagorean theorem, we have: $a^2 + b^2 = c^2 \rightarrow 8^2 + 15^2 = c^2 \rightarrow 64 + 225 = c^2 \rightarrow c = 17$, $sin x = \frac{opposite}{hypotenuse} = \frac{8}{17}$

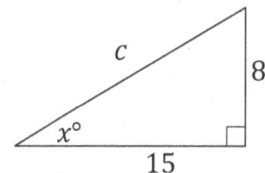

24) The answer is 1

The function $f(x)$ is undefined when the denominator of $\frac{1}{(x-3)^2 + 4(x-3) + 4}$ is equal to zero. The expression $(x - 3)^2 + 4(x - 3) + 4$ is a perfect square.

$(x-3)^2 + 4(x-3) + 4 = ((x-3)+2)^2$ which can be rewritten as $(x-1)^2$. The expression $(x-1)^2$ is equal to zero if and only if $x = 1$. Therefore, the value of x for which $f(x)$ is undefined is 1.

25) The answers are $0, -2, -4$

Frist factor the function: $f(x) = x^3 + 6x^2 + 8x = x(x+4)(x+2)$, To find the zeros, $f(x)$ should be zero. $f(x) = x(x+4)(x+2) = 0$, Therefore, the zeros are:

$x = 0$,

$(x+4) = 0 \Rightarrow x = -4$, $(x+2) = 0 \Rightarrow x = -2$

26) The answer is $\frac{12x-3}{x^3}$

First find a common denominator for both fractions in the expression $\frac{5}{x^2} + \frac{7x-3}{x^3}$.

The common denominator is x^3. Now, we can combine like terms into a single numerator over the denominator:

$$\frac{5x}{x^3} + \frac{7x-3}{x^3} = \frac{5x + (7x-3)}{x^3} = \frac{12x-3}{x^3}$$

27) The answer is 25

The angles on a straight line add up to 180 degrees. Then: $x + 25 + y + 2x + y = 180$

$\rightarrow 3x + 2y = 180 - 25$, $x = 35 \rightarrow 3(35) + 2y = 155 \rightarrow 2y = 155 - 105 = 50$

$\rightarrow y = 25$

28) The answer is $n^2 + 4n + 11$

$(3n^2 + 4n + 6) - (2n^2 - 5)$. Combine like terms together: $3n^2 - 2n^2 = n^2$

$6 - (-5) = 11$

Combine these terms into one expression to find the answer:

$(3n^2 + 4n + 6) - (2n^2 - 5) = n^2 + 4n + 11$

29)The answer is 28 *in*

Let x be the width of the rectangle. Use Pythagorean Theorem: $a^2 + b^2 = c^2$

$x^2 + 8^2 = 10^2 \Rightarrow x^2 + 64 = 100 \Rightarrow x^2 = 100 - 64 = 36 \Rightarrow x = 6$

The width of the rectangle is 6 inches. Then:

The perimeter of the rectangle $= 2\ (length + width) = 2\ (8 + 6) = 2\ (14) = 28\ in$

30)The answer is 1. 02

The initial deposit earns 2 percent interest compounded annually. Thus, at the end of year 1, the new value of the account is the initial deposit of $150 plus 2 percent of the initial deposit: $150 + \frac{2}{100}$ ($150) = $150(1.02)$.

Since the interest is compounded annually, the value at the end of each succeeding year is the sum of the previous year's value plus 2 percent of the previous year's value. This is equivalent to multiplying the previous year's value by 1.02. Thus, after 2 years, the value will be $150(1.02)\ (1.02) = \$(150)(1.02)^2$; and after 3 years, the value will be $(150)(1.02)^3$; and after n years, the value will be $(150)(1.02)^n$. Therefore, in the formula for the value for Sara's account after n years $(100)(x)^n$, the value of x is 1.02.

31)The answer is on the graph

To graph this line, we need to find two points. When x is zero the value of y is -5. And when y is 0 the value of x is 2.5.

$x = 0 \rightarrow y + 3 = 2(0 - 1) \rightarrow y = -5$

$y = 0 \rightarrow 0 + 3 = 2(x - 1) \rightarrow x = 2.5$

Now, we have two points: $(0, -5)$ and $(2.5, 0)$.

Find the points on the coordinate plane and graph the line.

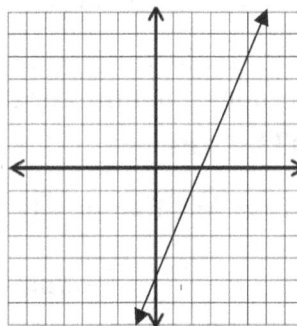

Remember that the slope of the line is 2.

32)The answer is $215 + 45i$

Use the FOIL (First, Out, In, Last) method to multiply two imaginary expressions:

$(-7)(-5) + (-7)(-15i) + (12i)(-5) + (12i)(-15i) = 35 + 105i - 60i - 180i^2$

Combine like terms $(+105i - 60i)$ and simplify:

$35 + 105i - 60i - 180i^2 = 35 - 45i - 180i^2$

$i^2 = -1,$

then: $35 - 45i - 180i^2 = 35 - 45i - 180(-1) = 35 - 45i + 180 = 215 + 45i$

33)The answer is $\log s + \frac{1}{2} \log t - 2 \log r$

Use logarithms rule: $\log_a x - \log_a y = \log_a \frac{x}{y}$. Then: $\log \frac{s\sqrt{t}}{r^2} = \log s\sqrt{t} - \log r^2$

Using the logarithms rule: $\log_a(x \cdot y) = \log_a x + \log_a y$, we get: $\log s\sqrt{t} = \log s + \log \sqrt{t}$

We know that $\sqrt{t} = t^{\frac{1}{2}}$. Then: $\log s + \log \sqrt{t} = \log s + \log \sqrt{t} = \log s + \log t^{\frac{1}{2}}$

Now, use logarithms rule: $\log_a x^n = n \log_a x$. Then: $\log t^{\frac{1}{2}} = \frac{1}{2} \log t$ and $\log r^2 = 2 \log r$

Finally: $\log(s\sqrt{t}) - \log(r^2) = \log s + \frac{1}{2} \log t - 2 \log r$

34)The answer is $2(x-6)(x+6)$ or $2x^2 - 72$

Find the factors of the denominators: $\frac{3x}{x^2-36} = \frac{3x}{(x-6)(x+6)}$ and $\frac{4}{2x-12} = \frac{4}{2(x-6)}$

Since the factor $(x-6)$ is common in both denominators, then, the least common denominator is: $2(x-6)(x+6) = 2x^2 - 72$

35)The answer is 3.9

To find AC, use sine B. Then:

$sine \; \theta = \frac{opposite}{hypotenuse}$. $sine \; 40° = \frac{AC}{6} \rightarrow 6 \times sine \; 40° = AC,$

Now use a calculator to find $sine \; 40°$. $sine \; 40° \approx 0.643 \rightarrow AC \approx 3.86$

ALEKS Mathematics Practice Test 2
Answers and Explanations

1) **The answer is $-54 + 12i$**

Use the FOIL (First, Out, In, Last) method to multiply two imaginary expressions: $(-3 + 9i)(3 + 5i) = -9 - 15i + 27i + 45i^2$

We know that: $i^2 = -1$. Then: $-9 - 15i + 27i + 45i^2 = -9 + 12i - 45 = -54 + 12i$

2) **The answer is $30x^5 u^7$**

Apply Exponent's rules: $x^a \times x^b = x^{a+b}$

Then: $3x^4 u^5 . 5u^2 . 2x = 30x^5 u^7$

3) **The answers are 0 and -5**

To solve for u, divide both sides by 3. Then: $3u^2 + 15u = 0 = \frac{3u^2}{3} + \frac{15u}{3} = \frac{0}{3} =$

$u^2 + 5u = 0$. Take the common factor u out: $u^2 + 5u = 0 = u(u + 5) = 0$

The product of u and $(u + 5)$ is zero. Therefore, one or both factors must be zero. Then: $u = 0$, or $u + 5 = 0 \rightarrow u = -5$

4) **The answer is $(2x - 5)(2x + 3)$**

Find the factors of the product of a and c (ac) that add to b. $ac = 4(-15) = -60$, $b = -4$

Write bx as a sum or difference using the factors from step 1. $-4x = -10x + 6x$

Divide the polynomial into 2 groups. $4x^2 - 10$ and $+6x - 15$.

Take the common factor out of both groups. Then: $2x(2x - 5) + 3(2x - 5)$

Factor out the common binomial factor.

$2x(2x - 5) + 3(2x - 5) = (2x - 5)(2x + 3)$

5) The answer is: Domain: all real numbers except -2, Range: all real numbers except zero

The function is not defined for $x = -2$, as this value would result is division by zero.

Hence the domain of $f(x)$ is "all real numbers except -2.

Range: No matter how large or small x becomes, $f(x)$ will never be equal to zero.

So, the range of $f(x)$ is "all real numbers except zero".

6) The answer is $x^3 - 21x - 20$

Use the FOIL method (First, Out, In, Last) and then combine like terms:

$(x - 5)(x^2 + 5x + 4) = x^3 + 5x^2 + 4x - 5x^2 - 25x - 20 = x^3 - 21x - 20$

7) The answer is -38

Use PEMDAS (order of operation):

$5 + 8 \times (-3) - [4 + 22 \times 5] \div 6 = 5 + 8 \times (-3) - [4 + 110] \div 6$

$= 5 + 8 \times (-3) - [114] \div 6 = 5 + (-24) - 19 = 5 + (-24) - 19 = 5 - 43 = -38$

8) The answer is $\dfrac{-x - 3}{2x^2 - 10}$

To simplify this rational expression, combine the fractions in the denominator.

Then: $\dfrac{\frac{1}{2} - \frac{x + 5}{4}}{\frac{x^2}{2} - \frac{5}{2}} = \dfrac{\frac{1}{2} - \frac{x + 5}{4}}{\frac{x^2 - 5}{2}}$, now simplify the fractions in the numerator:

$\dfrac{\frac{1}{2} - \frac{x + 5}{4}}{\frac{x^2 - 5}{2}} = \dfrac{\frac{2}{4} - \frac{x + 5}{4}}{\frac{x^2 - 5}{2}} = \dfrac{\frac{2 - x - 5}{4}}{\frac{x^2 - 5}{2}}$

Apply this fraction rule: $\dfrac{\frac{a}{b}}{\frac{c}{d}} \rightarrow \dfrac{a \times d}{b \times c}$. Then: $\dfrac{\frac{-x - 3}{4}}{\frac{x^2 - 5}{2}} = \dfrac{2(-x - 3)}{4(x^2 - 5)} = \dfrac{-2x - 6}{4x^2 - 20}$, divide both

numerator and denominator by 2. Then: $\dfrac{-2x - 6}{4x^2 - 20} = \dfrac{-x - 3}{2x^2 - 10}$

9) The answer is $-\sqrt{3}$

$$tan\frac{2\pi}{3} = \frac{sin\frac{2\pi}{3}}{cos\frac{2\pi}{3}} = \frac{\frac{\sqrt{3}}{2}}{-\frac{1}{2}} = -\sqrt{3}$$

10) The answer is $x = 15$

To solve for x, isolate the radical on one side of the equation.

Divide both sides by 4. Then: $4\sqrt{2x+6} = 24 \rightarrow \frac{4\sqrt{2x+6}}{4} = \frac{24}{4} \rightarrow \sqrt{2x+6} = 6$

Square both sides: $\left(\sqrt{(2x+6)}\right)^2 = 6^2$, Then: $2x + 6 = 36 \rightarrow 2x = 30 \rightarrow x = 15$

Substitute x by 15 in the original equation and check the answer:

$x = 15 \rightarrow 4\sqrt{2(15)+6} = 4\sqrt{36} = 4(6) = 24$

So, the value of 15 for x is correct.

11) The answer is $C(4,-7), r = \sqrt{3}$

The equation of a circle in standard form is: $(x - h)^2 + (y - k)^2 = r^2$, where the center is at: (h, k) and its radius is: r

Then, for the circle with equation $(x - 4)^2 + (y + 7)^2 = 3$, the center is at $(4, -7)$ and its radius is $\sqrt{3}$ $(r^2 = 3 \rightarrow r = \sqrt{3})$

12) The answer is $(x + 4)^2 + (y - 2)^2 = 1$

First, find the radius of the circle. The circumference of a circle $= 2\pi \Rightarrow$

circumference $= 2\pi r = 2\pi \Rightarrow r = 1$

The equation of a circle in the coordinate plane: $(x - h)^2 + (y - k)^2 = r^2 \Rightarrow$

Center: (h, k) and radius: r, center: $(-4, 2) \Rightarrow h = -4, k = 2$

Then, the equation of the circle is: $(x + 4)^2 + (y - 2)^2 = 1$

13) The answer is $x = 31$

The two angles are complementary angles (their sum is 90 degrees). Then:

$2x - 6 + (x + 3) = 90 \rightarrow 3x - 3 = 90 \rightarrow 3x = 93 \rightarrow x = 31$

14) The answer is $n = t^2 - m$

To solve for n, use cross multiplication: $\frac{1}{m+n} = \frac{1}{t^2} \rightarrow m + n = t^2 \rightarrow n = t^2 - m$

15) The answers are $485°$ and $-235°$

Coterminal angles are equal angles. To find a Coterminal of an angle, add or subtract 360 degrees (or 2π for radians) to the given angle. Then:

$125° + 360° = 485°$

$125° - 360° = -235°$

16) The answer is $x \geq 2$

To solve the inequality, bring the variable x to one side by adding or subtracting. Then: $8x - 4 \geq -2x + 16 \rightarrow 8x + 2x - 4 \geq -2x + 16 + 2x \rightarrow 10x - 4 + 4 \geq 16 + 4$

$\rightarrow 10x \geq 20 \rightarrow x \geq \frac{20}{10} \rightarrow x \geq 2$

Since the variable is greater than or equal to 2, then we need to find 2 on the number line and draw an closed circle on it. Then, draw an arrow to the right.

17) The answer is $(x - 4)(x - 3)$

To factor this trinomial expression, break the expression into groups. You need to find two numbers that their product is 12 and their sum is -7. (remember "Reverse FOIL": $x^2 + (b + a)x + ab = (x + a)(x + b)$). Those two numbers are -3 and -4. Then:

$x^2 - 7x + 12 = (x^2 - 3x) + (-4x + 12)$

Now factor out x from $x^2 - 3x : x(x - 3)$, and factor out -4 from

$-4x + 12: -4(x - 3)$; Then: $(x^2 - 3x) + (-4x + 12) = x(x - 3) - 4(x - 3)$

Now factor out like term: $(x - 3).\rightarrow x^2 - 7x + 12 = (x - 4)(x - 3)$

18)The answer is $x = -\dfrac{7}{13}$

Use distributive property to simplify $2(12x + 6)$ and $-2(x + 1)$. Then:

$2(12x + 6) = 24x + 12$ and $-2(x + 1) = -2x - 2$

Isolate the variable: $24x + 12 = -2x - 2$, subtract 12 from both sides.

$\rightarrow 24x + 12 - 12 = -2x - 2 - 12 \rightarrow 24x = -2x - 14$, add $2x$ to both sides:

$\rightarrow 24x + 2x = -2x - 14 + 2x \rightarrow 26x = -14$. Divide both sides by 26:

$\rightarrow \dfrac{26x}{26} = \dfrac{-14}{26} \rightarrow x = -\dfrac{7}{13}$

19)The answer is 5

Use distance of two points formula: $d = \sqrt{(x_A - x_B)^2 + (y - y_B)^2} \rightarrow$

$d = \sqrt{(1 - (-2))^2 + (3 - 7)^2} = \sqrt{(3)^2 + (-4)^2} = \sqrt{9 + 16} = \sqrt{25} = 5$

The distance between the two points on the coordinate plane is 5 units.

20)The answer is: Solution $x < -2$ or $x > 8$, Interval Notation: $(-\infty, -2) \cup (8, \infty)$

Factor the numerator: $\dfrac{6x+12}{x-8} > 0 \rightarrow \dfrac{6(x+2)}{x-8} > 0$

Divide both sides by 6: $\rightarrow \dfrac{\frac{6(x+2)}{x-8}}{6} > \dfrac{0}{6} \rightarrow \dfrac{x+2}{x+8} > 0 \rightarrow$ Find the signs of the factors $\dfrac{x+2}{x+8}$.

Plug in some values of x and check the solutions. Only $x < -2$ or $x > 8$ work in the inequality. The interval notation: $(-\infty, -2) \cup (8, \infty)$

21) The answer is 31

Find the sum of five numbers.

$average = \frac{sum\ of\ terms}{number\ of\ terms} \Rightarrow 26 = \frac{sum\ of\ 5\ numbers}{5} \Rightarrow sum\ of\ 5\ numbers = 26 \times 5 = 130$

The sum of 5 numbers is 130. If a sixth number 56 is added, then the sum of 6 numbers is $130 + 56 = 186$. The new average is: $\frac{sum\ of\ 6\ numbers}{6} = \frac{186}{6} = 31$

22) The answer is $\frac{x^2-2x-24}{2x-2}$

To rewrite $\frac{1}{\frac{1}{x-6}+\frac{1}{x+4}}$, first simplify $\frac{1}{x-6}+\frac{1}{x+4}$.

$\frac{1}{x-6}+\frac{1}{x+4} = \frac{1(x+4)}{(x-6)(x+4)} + \frac{1(x-5)}{(x+4)(x-6)} = \frac{(x+4)+(x-6)}{(x+4)(x-6)}$

Then: $\frac{1}{\frac{1}{x-6}+\frac{1}{x+4}} = \frac{1}{\frac{(x+4)+(x-6)}{(x+4)(x-6)}} = \frac{(x-6)(x+4)}{(x-6)+(x+4)}$. (Remember, $\frac{1}{\frac{1}{x}} = x$)

Now, simplify denominator by combining like terms and the numerator by using the FOIL method. Then:

$$\frac{(x-6)(x+4)}{(x-6)+(x+4)} = \frac{x^2-2x-24}{2x-2}$$

23) The answer is $y = 2x^2 - 4x + 4$

In order to figure out what the equation of the graph is, fist find the vertex. From the graph we can determine that the vertex is at $(1,2)$. We can use vertex form to solve for the equation of this graph. Recall vertex form, $y = a(x-h)^2 + k$, where h is the x coordinate of the vertex, and k is the y coordinate of the vertex. Plugging in our values, you get $= a(x-1)^2 + 2$, To solve for a, we need to pick a point on the graph and plug it into the equation. Let's pick $(-1, 10)$, $10 = a(-1-1)^2 + 2$

$10 = a(-2)^2 + 2$, $10 = 4a + 2$, $8 = 4a$, $a = 2$

Now the equation is: $y = 2(x - 1)^2 + 2$

Let's expand this, $y = 2(x^2 - 2x + 1) + 2$, $y = 2x^2 - 4x + 2 + 2 \rightarrow y = 2x^2 - 4x + 4$

24)The answer is $\frac{2x^2 + xy - 2y^2}{x - y}$

To simplify this expression, take the common factors out. Then:

$$\frac{4x^3 + 2x^2y - 4xy^2}{2x^2 - 2xy} = \frac{2x(2x^2 + xy - 2y^2)}{2x(x - y)} \rightarrow \frac{2x^2 + xy - 2y^2}{x - y}$$

25)The answer is $(-1 + \sqrt{6})$ **and** $(-1 - \sqrt{6})$

Use quadratic formula: $x_{1,2} = \frac{-b \pm \sqrt{b^2 - 4ac}}{2a}$

The quadratic equation in standard form is: $ax^2 + bx + c = 0$

For the equation: $x^2 + 2x - 5 = 0 \Rightarrow$ Then: $a = 1$, $b = 2$ and $c = -5$

$x = \frac{-2 + \sqrt{2^2 - 4 \times 1 \times (-5)}}{2 \times 1} = \frac{-2 + \sqrt{4 - (-20)}}{2} = \frac{-2 + \sqrt{24}}{2} = \frac{-2 + \sqrt{6 \times 4}}{2} = \frac{-2 + 2\sqrt{6}}{2} = \frac{2(-1 + \sqrt{6})}{2} =$

$-1 + \sqrt{6}$ or $x = \frac{-2 - \sqrt{2^2 - 4 \times 1 \times (-5)}}{2 \times 1} = -1 - \sqrt{6}$

26)The answer is on the graph

To draw the graph of $y < -2x - 2$*, you first need to graph the line:* $y = -2x - 2$

Since there is a less than ($\leq$) sign, draw a solid line.

The slope is -2 and y-intercept is -2.

Then, choose a testing point and substitute the value of x and y from that point into the inequality. The easiest point to test is the origin: $(0, 0)$

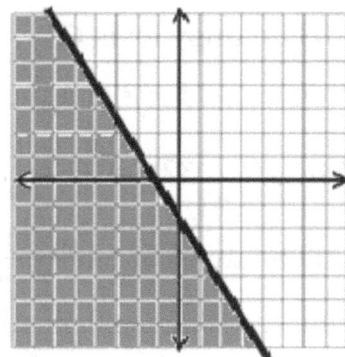

$(0,0) \rightarrow y < -2x - 2 \rightarrow 0 < -2(0) - 2 \rightarrow 0 < -2$

This is incorrect! 0 is not less than -2. So, the left side of the line is the solution of this inequality.

27) The answer is 570

This is a simple matter of substituting values for variables. We are given that the 50 cars were washed today, therefore we can substitute that for a. Giving us the expression $P(x) = \frac{40(50)-500}{50} + b$. We are also given that the profit was \$600, which we can substitute for $P(a)$. Which gives us the equation $600 = \frac{40(50)-500}{50} + b$

Simplifying the fraction gives us the equation $600 = 30 + b$

And subtracting both sides of the equation by 30 gives us $b = 570$, which is the answer.

28) The answer is 32

Logarithm is another way of writing exponent. $log_b{}^y = x$ is equivalent to $y = b^x$.

Rewrite the logarithm in exponent form: $log_2 x = 5 \rightarrow 2^5 = x \rightarrow x = 32$

29) The answer is 4

To solve this problem first solve the equation for c. $\frac{c}{b} = 2$

Multiply by b on both sides. Then: $b \times \frac{c}{b} = 2 \times b \rightarrow c = 2b$. Now to calculate $\frac{8b}{c}$,

substitute the value for c into the denominator and simplify. $\frac{8b}{c} = \frac{8b}{2b} = \frac{8}{2} = 4$

30) The answer is 60

Plug in 140 for F and then solve for C.

$$C = \frac{5}{9}(F - 32) \Rightarrow C = \frac{5}{9}(140 - 32) \Rightarrow C = \frac{5}{9}(108) = 60$$

31) The answer is $\frac{2x^2-23x-56}{x^2-64}$

Find the common denominator. It's $(x + 8)(x - 8)$. Then:

$$\frac{2x}{x+8} - \frac{7}{x-8} = \frac{2x(x-8)}{(x+8)(x-8)} - \frac{7(x+8)}{(x-8)(x+8)} = \frac{2x^2-16x-(7x+56)}{(x+8)(x-8)} = \frac{2x^2-23x-56}{x^2-64}$$

32)The answer is $y = 0$

In a rational function, if the denominator has a bigger degree than the numerator, the horizontal asymptote is the x-axes or the line $y = 0$. In the function $f(x) = \frac{x+3}{x^2+1}$, the degree of numerator is 1 (x to the power of 1) and the degree of the denominator is 2 (x to the power of 2). Then, the horizontal asymptote is the line $y = 0$.

33)The answer is 44

To find the value of x, use the cosine on the angle x:

$$\cos\theta = \frac{adjacent}{hypotenuse} \rightarrow \cos x = \frac{10}{14} = \frac{5}{7}$$

Use a calculator to find inverse cosine: $cos^{-1}\left(\frac{5}{7}\right) = 44.41° \approx 44°$

Then: $x = 44$

34)The answer is $\frac{5\pi}{6}$

Use this formula: Radians = Degrees $\times \frac{\pi}{180}$

Radians $= 150 \times \frac{\pi}{180} = \frac{150\pi}{180} = \frac{5\pi}{6}$

35)The answer is 6.08×10^7

Convert the second number to have the same power of 10.

$3.2 \times 10^6 = 0.32 \times 10^7$

Now, two numbers have the same power of 10. Factor 10^7 out.

$6.4 \times 10^7 - 0.32 \times 10^7 = (6.4 - 0.32) \times 10^7$

Subtract: $6.4 - 0.32 = 6.08$. Then: $(6.4 - 0.32) \times 10^7 = 6.08 \times 10^7$

... So Much More Online!

Effortless Math Online ALEKS Math Center offers a complete study program, including the following:

✓ Step-by-step instructions on how to prepare for the ALEKS Math test

✓ Numerous ALEKS Math worksheets to help you measure your math skills

✓ Complete list of ALEKS Math formulas

✓ Video lessons for ALEKS Math topics

✓ Full-length ALEKS Math practice tests

✓ And much more...

No Registration Required.

Receive the PDF version of this book or get another FREE book!

Thank you for using our Book!

Do you LOVE this book?

Then, you can get the PDF version of this book or another book absolutely FREE!

Please email us at:

info@EffortlessMath.com

for details.

Author's Final Note

I hope you enjoyed reading this book. You've made it through the book! Great job!

First of all, thank you for purchasing this study guide. I know you could have picked any number of books to help you prepare for your ALEKS Math test, but you picked this book and for that I am extremely grateful.

It took me years to write this study guide for the ALEKS Math because I wanted to prepare a comprehensive ALEKS Math study guide to help test takers make the most effective use of their valuable time while preparing for the test.

After teaching and tutoring math courses for over a decade, I've gathered my personal notes and lessons to develop this study guide. It is my greatest hope that the lessons in this book could help you prepare for your test successfully.

If you have any questions, please contact me at reza@effortlessmath.com and I will be glad to assist. Your feedback will help me to greatly improve the quality of my books in the future and make this book even better. Furthermore, I expect that I have made a few minor errors somewhere in this study guide. If you think this to be the case, please let me know so I can fix the issue as soon as possible.

If you enjoyed this book and found some benefit in reading this, I'd like to hear from you and hope that you could take a quick minute to post a review on the book's Amazon page. To leave your valuable feedback, please visit: amzn.to/3QdNlHR

Or scan this QR code.

I personally go over every single review, to make sure my books really are reaching out and helping students and test takers. Please help me help ALEKS Math test takers, by leaving a review!

I wish you all the best in your future success!

Reza Nazari

Math teacher and author

Made in United States
Cleveland, OH
10 July 2025